Contents

Chapter 9 Industrial and commercial switchgear

Chapter 10 Electromagnetic compatibility

Cooperating organisations

The Institution of Engineering and Technology acknowledges the contribution made by the following representatives of organisations in the preparation of this publication:

Association of Manufacturers of Domestic Electrical Appliances
S. MacConnacher

BEAMA (The British Electrotechnical and Allied Manufacturers Association)
P. Galbraith, M. Mullins, P. Sayer

Department of Trade and Industry
G. Scott

Electrical Contractors' Association
D. Locke, R. Lovegrove

Electrical Contractors' Association of Scotland
D. Millar

Health and Safety Executive
N. Gove

Institution of Engineering and Technology
M. Coles, G. Cronshaw, S. Stewart

John Lewis
W. Wright

Lighting Association
K. R. Kearney

National Inspection Council for Electrical Installation Contracting
M. Clark

Office of the Deputy Prime Minister
K. Bromley

C. S. Todd and Associates Ltd
C. Todd

West Green Associates
G. Willard

Author
J. Ware

Electrical Maintenance

The need for maintenance

1.1 General

Electrical maintenance is carried out for four basic reasons:

1 to prevent danger
2 to reduce unit cost
3 to keep a facility in operation (reliability)
4 to prevent pollution of the environment.

The need to prevent danger is enforced by much legislation. Cost and reliability are matters to be decided probably on managerial criteria only. Protection of the environment is supported by legislation.

Balancing the cost of maintenance against the cost of breakdown generally involves managerial decisions that may be divorced from safety considerations. These matters are discussed in Chapter 2. This publication deals with electrical maintenance. However, many electrical maintenance activities produce waste (oil, old lamps, etc.) the impact of which upon the environment needs consideration. This is discussed in Chapter 12, together with the relevant legislation.

1.2 Safety legislation, Codes of Practice and guidance

Apart from the common law general duty of care of everyone for his neighbour there is specific legislation with respect to safety, the most fundamental being the *Health and Safety at Work etc. Act 1974*. This Act empowers the Secretary of State to make regulations, generally referred to as Health and Safety Regulations. A number of these Health and Safety Regulations are discussed where they are relevant to electrical maintenance. The *Health and Safety at Work etc. Act* empowers the Health and Safety Commission to approve and issue codes of practice. These codes of practice may be prepared by the Commission or by others. The consent of the Secretary of State is required before the Commission approves a code of practice. The failure on the part of any person to observe a provision of an approved code of practice does not of itself render him liable to any civil or criminal proceedings, but the breach of a provision of a code of practice could be used as evidence of non-compliance with a regulation. The approved codes of practice issued by the Health and Safety Commission consequently have a special place in the Health and Safety at Work legislation.

1.2.1 *Health and Safety at Work etc. Act 1974*
The *Health and Safety at Work etc. Act 1974* is comprehensive and is concerned with health, safety and welfare at work, the control of dangerous substances and certain emissions into the atmosphere.

Sections 2, 3 and 4 of the *Health and Safety at Work etc. Act 1974* put a duty of care upon the employer, the employee and the self-employed to ensure the health, safety and welfare at work of all persons, employees and others using the work premises.

Section 6 imposes a duty on any person who designs or manufactures any article for use at work to take such steps as are necessary to ensure that there will be available, in connection with the use of the article at work, adequate information on the use for which the article has been designed and tested. It also requires that advice be given on any conditions necessary to ensure that, when put to use, it will be safe, and without risks to health. This implies a general duty that persons supplying electrical installations and electrical equipment must ensure not only that the equipment they supply is safe and adequately tested, but that adequate information is provided for it to be maintained in a safe condition.

The Act empowers the Secretary of State to make regulations, and those most relevant to electrical installation work are:

▸ the *Electricity at Work Regulations 1989*
▸ the *Management of Health and Safety at Work Regulations 1999*
▸ *Construction (Design and Management) Regulations 1994*
▸ *Provision and Use of Work Equipment Regulations 1998*
▸ *Personal Protective Equipment at Work Regulations 1992*.

Lack of knowledge of a regulation is no defence in law.

Further information on legislation relevant to persons engaged in electrical installations is given in the IEE publication *Commentary on the IEE Wiring Regulations 16th edition*.

1.2.2 The *Electricity at Work Regulations 1989*

The *Electricity at Work Regulations 1989* (EWR) (Statutory Instrument No. 635) impose duties on every employer, every employee and every self-employed person to ensure that the safety requirements of the regulations are met. The EWR are enacted to provide for the electrical safety of persons in the workplace. Regulation 4 of the EWR deals with electrical systems, work activities and protective equipment and requires that all electrical systems be designed, installed and maintained in use so as to prevent danger. The regulation states, specifically:

1 All systems shall at all times be of such construction as to prevent, so far as is reasonably practicable, danger.
2 As may be necessary to prevent danger, all systems shall be maintained so as to prevent, so far as is reasonably practicable, such danger.
3 Every work activity, including operation, use and maintenance of a system and work near a system, shall be carried out in such a manner as not to give rise, so far as is reasonably practicable, to danger.
4 Any equipment provided under these Regulations for the purpose of protecting persons at work on or near electrical equipment shall be suitable for the use for which it is provided, be maintained in a condition suitable for that use, and be properly used.

Memorandum of guidance on the
Electricity at Work Regulations 1989

GUIDANCE
ON REGULATIONS

▸ **Figure 1.1** HSR25:
Memorandum of Guidance on the Electricity at Work Regulations

In practice, therefore, all electrical systems must be designed, installed, maintained and used so as to prevent danger. The term 'system' means electrical system and includes all the electrical equipment that is, or may be, connected to a common source of electrical energy. This includes the fixed installation and all equipment that may be supplied from it, and as such also embraces equipment supplied via a plug and socket.

The *Memorandum of Guidance on the Electricity at Work Regulations* published by the Health and Safety Executive (publication HSR25) provides guidance on all aspects of the EWR. With respect to the design and installation of electrical systems it says that

'The IEE Wiring Regulations (BS 7671) is a code of practice which is widely recognised and accepted in the UK compliance with which is likely to achieve compliance with the relevant aspects of the 1989 Regulations'.

However, the scope of the *Electricity at Work Regulations* (EWR) is much wider than that of BS 7671: *Requirements for Electrical Installations* (the *IEE Wiring Regulations*). The EWR are also concerned with maintenance of the installation, training and competence of staff, good working practices and suitable equipment. The EWR require:

▹ installations to be constructed so as to be safe; regulations 4(1), 5, 6, 7, 8, 9, 10, 11, 12
▹ installations to be maintained so as to be safe; regulation 4(2)
▹ associated work to be carried out safely; regulations 4(3), 13, 14, 15
▹ work equipment provided to be suitable for the purpose; regulation 4(4) and
▹ persons to be competent; regulation 16.

The EWR deal principally with electrical installation and fixed electrical equipment. Information concerning the maintenance of portable equipment, appliances, industrial, business and office equipment is available in the IEE publication *In-Service Inspection and Testing of Electrical Equipment*.

1.2.3 *Management of Health and Safety at Work Regulations 1999*

The prime requirement of the *Management of Health and Safety at Work Regulations* is that:

Every employer shall make a suitable and sufficient assessment of:

a the risks to the health and safety of his employees to which they are exposed while they are at work
b the risks to the health and safety of persons not in his employment arising out of or in connection with the conduct by him of his undertaking.

▹ **Figure 1.2** Approved Code of Practice L21

The *Management of Health and Safety at Work Regulations* apply to all risks to the health and safety of persons whereas the *Electricity at Work Regulations* deal with electrical systems and work on, associated with or near such systems. The *Management Regulations* introduce the concept of risk assessment in the broadest sense. The level of detail required within the risk assessment is to be proportional to the hazard. Significant findings are required to be recorded, together with the control measures taken.

These regulations have requirements for health and safety training. Work entrusted to employees should take into account their capabilities, and adequate health and safety training is required to be provided (regulation 13). No person must be engaged in any work activity where technical knowledge or experience is necessary to prevent danger or, where appropriate, injury, unless they possess such knowledge or experience or are under a degree of supervision as may be appropriate having regard to the nature of the work (regulation 16 of the *Electricity at Work Regulations* refers).

As well as imposing duties upon the employer, the Regulations impose duties on employees. Every employee is required to use machinery, equipment etc. provided to him by his employer in accordance with any training he has received and any instructions provided to him (regulation 14). There is also a duty upon every employee to inform his employer where he has concerns regarding the safety of fellow employees, both with respect to equipment provided and training. This is a particularly relevant requirement for those with general responsibilities for maintenance, including those carrying out the maintenance work. If any person has concerns regarding his ability to carry out work he must bring this to the attention of his employer.

The Health and Safety Executive has issued an Approved Code of Practice L21, for the *Management of Health and Safety at Work Regulations 1999*. As this is an Approved Code of Practice, non-compliance with any recommendations of the Code can be used as evidence of non-compliance with the *Management of Health and Safety at Work Regulations*. In all but the most straightforward businesses, it will be necessary to obtain a copy of the Code and implement all the relevant recommendations.

A prime requirement of the *Management of Health and Safety at Work Regulations* is that every employer shall make a suitable and sufficient assessment of the risks to the health and safety of his employees to which they are exposed while at work and the risks to health and safety to persons not in his employment arising out of, or in connection with, the work of his undertaking. The Code of Practice provides guidance on the assessment of these risks.

A structured approach, as outlined below, may assist in maintenance management.

Policy statement

As an indication of the senior management or directors' commitment to health and safety at work, it is usual for a policy statement to be drafted and signed by the senior members of the company.

Responsibilities

The particular responsibilities of individuals within the company for safety must be identified, including who has final responsibility for implementing the safety policy of the company.

Determination of risks

Using the guidance given by the Approved Code of Practice the risks to the health and safety of employees and others associated with the work of the company need to be assessed. After carrying out the risk assessment, decisions have to be made about the control measures necessary to manage the risks.

Control measures

Control measures must be taken to ensure that the identified risks are properly handled. This may require:

1 purchase of equipment
2 changes of procedure
3 issuing of instructions
4 training of staff
5 provision of equipment to staff
6 control/feedback facilities.

Safety instructions

Safety instructions, written and formalised, are an effective management tool for the implementation of control measures. They reduce uncertainty about responsibilities, may detail the control measures to be taken and can be used as a reference as well as an instruction to staff on their own responsibilities and the procedures to be followed. They also remind management of their responsibilities to ensure that instructions are complied with and that training and updating is provided.

Training

The safety instructions will have specified the level of training necessary for staff and this will have been determined before the safety instructions are written up. Employees will normally have been trained as necessary, but the issue of safety instructions will enable this to be checked, both by management and the staff themselves, with further training arranged as required including training in the use of necessary equipment.

Equipment

The safety instructions will make the staff aware of the equipment they should be using, of their responsibility to use equipment provided for them and their responsibilities to keep such equipment in good order, advising supervisors of any deficiencies in the equipment, training and instructions.

1.2.4 Workplace (Health, Safety and Welfare) Regulations 1992

The *Workplace (Health, Safety and Welfare) Regulations 1992* require that every employer shall ensure that the workplace equipment, devices and systems are maintained. This includes keeping the equipment, devices and systems in an efficient state, in efficient working order and in good repair. Where appropriate, the equipment, devices and systems shall be subject to a suitable system of maintenance.

The scope of the *Workplace (Health, Safety and Welfare) Regulations* is somewhat different to that of the *Electricity at Work Regulations*. The *Electricity at Work Regulations* are basically concerned with ensuring an electrical installation is in a safe condition, and ensuring work performed on an electrical installation is done in a safe manner. They do not deal with the consequences of maloperation of the electrical system. However, the *Workplace Regulations* are concerned with the consequences of equipment and system failures. For example, although a malfunctioning emergency lighting system may not in itself be an electrical hazard, there is a potential hazard if there is no emergency lighting. These regulations impose maintenance regimes upon such systems as emergency lighting, fire alarms, powered doors, escalators and moving walkways that have electrical power supplies. The regulations are not limited to electrical systems but also include equipment such as fencing, equipment used for window cleaning, devices to limit the opening of windows etc. The approved code of practice to the *Workplace Regulations* states that the maintenance of work electrical equipment and electrical systems is also addressed in other regulations. Electrical systems are clearly well addressed in the *Electricity at Work Regulations*, and the maintenance of work equipment in the *Provision and Use of Work Equipment Regulations 1992*.

> **Figure 1.3** Approved Code of Practice L24

1.2.5 Construction (Design and Management) Regulations 1994

The *Construction (Design and Management) Regulations 1994* (CDM) and the Approved Code of Practice L54 apply generally to construction work (regulation 3). The basic requirement is that design and construction must take account of health and safety aspects in the construction phase of the work and during any subsequent maintenance. The Regulations also have requirements regarding the cleaning, maintaining and repairing of the construction at any time, including after construction is completed, i.e. during use and during demolition of the construction.

There is a requirement to provide reasonably foreseeable information necessary for the health and safety of persons who will carry out maintenance, repairs and cleaning in the future. Consequently, persons with such responsibilities for a building should request access to the health and safety file prepared for the construction to see if there are any particular problems associated with maintenance and repair, including electrical matters.

Figure 1.4 HSE Guidance on Regulations L22

Figure 1.5 HSE Guidance on Regulations L25

Figure 1.6 HSE Guidance on Regulations L23

1.2.6 *Provision and Use of Work Equipment Regulations 1998*

The *Provision and Use of Work Equipment Regulations* (PUWER) require that work equipment including an electrical installation is so constructed or adapted as to be suitable for the purpose for which it is provided. Equipment must be inspected after installation and before use, at suitable intervals and after exceptional events, e.g. circuit-breaker operation under fault conditions. Potentially dangerous machinery must be guarded. Where necessary, records must be maintained and adequate training must be given to operators.

1.2.7 *Personal Protective Equipment at Work Regulations 1992*

The *Personal Protective Equipment at Work Regulations* require the employer to ensure that suitable personal protective equipment is provided to his employees, as may be necessary. The equipment must take account of the risks, the environmental conditions, ergonomic requirements and the state of health of the person or persons. The equipment must fit and it must comply with any appropriate provisions or standards.

The regulations require that an assessment be made to ensure that the protective equipment is suitable. Purchasers should be looking for approval body and CE marking to demonstrate compliance with appropriate standards. The relevance of the approval mark to the use must be confirmed. The *Notes of Guidance* published by the HSE make the important note that the initial selection is only a first stage in a continuing programme concerned with the proper selection, maintenance and use of the protective equipment. Training and supervision of users must be appropriate to the equipment. The employer is required to ensure that the employee has sufficient information, instruction and training to use the equipment provided and knows the limits of the protection provided by the equipment, and any supplementary action that might be necessary in particular circumstances. For example, in the selection of safety goggles, the employee must be made aware of what the goggles are protecting against, as of course one pair may not be suitable for all the potential risks.

There is a requirement upon employees to use protective equipment provided in accordance with training and instruction received.

Employees are required to report the loss of such equipment. The requirements of this legislation do mean that proper records should be kept of protective equipment and a suitable procedure be set up for checking that employees still have the equipment necessary and that it is in good order. This does not in any way reduce the duty of the employee to advise the employer of any defects or deficiencies in the equipment or the training that he or she has received.

1.2.8 *Manual Handling Operations Regulations 1992*

Employers are required, so far as is reasonably practicable, to avoid the need for employees to undertake any manual handling operations which involve the risk of their being injured. Where this is not practicable, employers are obliged to make a suitable and sufficient assessment of all such manual handling operations and take appropriate steps to reduce the risk of injury to the lowest level reasonably practicable and take appropriate steps to give employees information on the weight of each load and information on the heaviest side of any load whose centre of gravity is not positioned centrally.

Assessments should be reviewed by employers.

The manual handling guidance L23 published by the Health and Safety Executive provides general guidance on manual handling operations and includes examples of assessment checklists.

Further details are given in Chapter 2 of this publication.

1.2.9 *Health and Safety (Display Screen Equipment) Regulations 2002*

The *Health and Safety (Display Screen Equipment) Regulations 2002* place a requirement upon employers to ensure that workstations (desks with computer screens and keyboards) meet the requirements laid down in schedules to the regulations.

An employer is required, at the request of an employee, to provide an appropriate eyesight test for users of workstations or persons who are about to become users. There is a further requirement that the employer provides adequate training and information. These regulations are discussed further in Chapter 2.

1.2.10 The *Health and Safety (Safety Signs and Signals) Regulations 1996*

These regulations are discussed in detail in Chapter 2.

▶ **Figure 1.6** HSE Guidance Note L26

1.2.11 The *Fire Precautions (Workplace) Regulations 1997 (as amended)*

The Home Office and Scottish Office publish a guidance document *Fire Safety: an Employer's Guide* (ISBN 0 1134 122 90), providing information for employers about the *Fire Precautions (Workplace) Regulations 1997*.

The *Fire Precautions (Workplace) Regulations* require every employer to safeguard the safety of employees in the case of fire. They are required, where necessary, to equip the workplace with:

▶ fire fighting equipment
▶ fire detection and alarm systems.

It is a requirement that any non-automatic fire fighting equipment shall be easily accessible, simple to use and indicated by signs. The employer is required to take measures for fire fighting appropriate for the size and nature of the activity, number of employees etc., to nominate a person to implement these measures and to ensure that the arrangements are complied with.

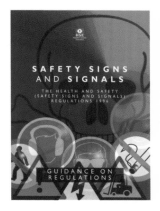

▶ **Figure 1.7** HSE Guidance Note L64

Where necessary to safeguard the safety of employees, arrangements must be made to ensure:

▶ routes to exits and the exits themselves are kept clear
▶ emergency routes lead to a place of safety
▶ it is possible to evacuate the workforce quickly and safely
▶ doors open in the direction of escape (sliding and revolving doors should not be the means of escape)
▶ emergency doors are not locked or fastened
▶ emergency routes and exits are indicated by signs
▶ emergency routes and exit signs are illuminated.

1.2.12 The *Building Regulations 2000, Statutory Instrument 2000 No. 2531*

The *Building Regulations 2000* (as amended) are implemented under powers provided in the *Building Act 1984*. They apply in England and Wales, but do not apply in Scotland or Northern Ireland. In Scotland the requirements of the *Building (Scotland) Regulations (2004)* apply. In Northern Ireland, the *Building Regulations (Northern Ireland) 2000* apply.

The purpose of the *Building Regulations* is to provide for the health and safety of people in and around buildings and also provide for issues including energy conservation, building access and building use.

In the *Building Regulations* 'building work' means any of the following:

▶ the erection or extension of a building
▶ the provision or extension of a controlled service or fitting in or in connection with a building
▶ the material alteration of a building, or a controlled service or fitting
▶ material change of use of the building (work required by regulation 6)
▶ the insertion of insulating material into the cavity wall of a building
▶ work involving the underpinning of a building.

Regulation 4 of the *Building Regulations 2000* requires:

1 Building work shall be carried out so that:
▶ it complies with the applicable requirements contained in Schedule 1 (Parts A to P)
▶ in complying with any such requirement there is no failure to comply with any other such requirement.
2 Building work shall be carried out so that, after it has been completed:
▶ any building which is extended or to which a material alteration is made, or
▶ any building in which, or in connection with which, a controlled service or fitting is provided, extended or materially altered, or
▶ any controlled service or fitting.

complies with the applicable requirements of Schedule 1 (Parts A to P) or, where it does not comply with any such requirement, is no more unsatisfactory in relation to that requirement than before the work was carried out.

The responsibility for compliance with the *Building Regulations* rests with the person carrying out the building work. The person carrying out the electrical installation is personally responsible for ensuring compliance. Persons employing others to carry out work should confirm who has the responsibility for compliance with the *Building Regulations*. Should there be non-compliance, the owner of the building is likely to be served with an enforcement notice.

The Building (Amendment No. 2) Regulations 2004 Statutory Instrument 2004 No. 1808

Amendment No. 2 to the *Building Regulations* introduced Part P into the Regulations.

An electrical installation is defined in the amendment: 'Electrical installation means fixed electrical cables or fixed electrical equipment located on the consumer's side of the electricity supply meter.'

For the purposes of providing practical guidance with respect to the requirements of the *Building Regulations 2000* for England and Wales, the Secretary of State has issued a series of Approved Documents. The Approved Documents include Part P: Electrical safety. The list of Approved Documents is given in Table 1.1.

The documents in Table 1.1 may be downloaded from the Office of the Deputy Prime Minister (ODPM) website: www.odpm.gov.uk.

Approved document	Title	Requirements relevant to an electrician working in a dwelling
Part A	Structure	depth of chases in walls, and size of holes and notches in floor and roof joists
Part B	Fire safety	fire safety of certain electrical installations; provision of fire alarm and fire detection systems; fire resistance of penetrations through floors and walls
Part C	Site preparation and resistance to moisture	moisture resistance of cable penetrations through external walls
Part E	Resistance to the passage of sound	penetrations through floors and walls
Part F	Ventilation	ventilation rates for dwellings
Part L	Conservation of fuel and power	energy efficient lighting
Part M	Access to and use of buildings	heights of switches and socket-outlets
Part P	Electrical safety	electrical safety issues

In addition to the above requirements, electricians responsible for work within the scope of Part P of the *Building Regulations* may also be responsible for ensuring the requirements of other parts of the *Building Regulations* are met, where relevant, particularly if there are no other parties involved with the work.

▶ **Table 1.1** List of approved documents that give practical guidance to the *Building Regulations*

The *Building Regulations* (regulation 4(2)) require that, on completion of the work, the building should be no worse in terms of the level of compliance with the other applicable Parts of Schedule 1 to the *Building Regulations*, including Parts A, B, C, E, F, L and M. As an example, consider the case where a recessed luminaire is to be fitted on the ground floor of a two-storey dwelling. Cutting a hole in the ceiling for the luminaire may well degrade the fire and sound penetration characteristics of the floor (Approved Document B (*Fire safety*) and Approved Document E (*Sound penetration*) refer). A second example is where an electrician chases walls and drills joists for the installation of accessories and cables. The building structure must be no more unsatisfactory than before the work was carried out (Approved Document A refers).

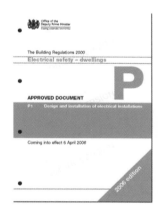

Approved Document P: Fixed Installations in Dwellings

The requirements of Part P of the *Building Regulations* and the limits on their application are stated in Table 1.2.

▶ **Figure 1.8** Approved Document P

Part P applies to electrical installations in buildings or parts of buildings comprising:

▶ dwelling houses and flats
▶ dwellings and business premises that have a common supply – for example, shops and public houses with a flat above
▶ common access areas in blocks of flats, such as corridors and staircases
▶ shared amenities of blocks of flats, such as laundries and gymnasiums
▶ land associated with the buildings – for example Part P applies to fixed lighting and pond pumps in gardens
▶ outbuildings such as sheds, detached garages and greenhouses.

Part P applies also to parts of the above electrical installations.

Requirement	Limits on application
Part P Electrical Safety Design, installation, inspection and testing P1 – Reasonable provision shall be made in the design, installation, inspection and testing of electrical installations in order to protect persons from fire or injury. Provision of information P2 – Sufficient information shall be provided so that persons wishing to operate, maintain or alter an electrical installation can do so with reasonable safety.	The requirements of this Part apply only to electrical installations that are intended to operate at low or extra-low voltage and are: ▶ in a dwelling ▶ in the common parts of a building serving one or more dwellings, but excluding power supplies to lifts ▶ in a building that receives its electricity from a source located within or shared with a dwelling ▶ in a garden or in or on land associated with a building where the electricity is from a source located within or shared with a dwelling.

▶ **Table 1.2**
Requirements of Part P of the *Building Regulations*

Compliance with Part P

In the Secretary of State's view, the requirements of Part P will be met by adherence to the fundamental principles for achieving safety given in Chapter 13 of BS 7671: 2001 *Requirements for Electrical Installations*.

To achieve the requirements of Chapter 13, electrical installations in dwellings etc. must be:

▶ designed and installed to afford appropriate protection against mechanical and thermal damage, and so that they do not present electric shock and fire hazards to people, and
▶ suitably inspected and tested to verify that they meet the relevant equipment and installation standards.

Chapter 13 of BS 7671 is met by complying with Parts 3 to 7 of BS 7671, except as allowed by regulations 120-01-03 and 120-02:

▶ Any intended departure from these parts (Parts 3 to 7) requires special consideration by the designer of the installation and is to be noted on the Electrical Installation Certificate specified in Part 7 (regulation 120-01-03).
▶ Where the use of a new material or invention leads to departures from the regulations, the resulting degree of safety of the installation is to be not less than that obtained by compliance with the regulations. Such use is to be noted on the Electrical Installation Certificate specified in Part 7 (regulation 120-02-01).

1.2.13 BS 7671: 2001 *Requirements for Electrical Installations*

Since its publication on 10 May 1991, the 16th edition of the *IEE Wiring Regulations* has been adopted by the British Standards Institution as the United Kingdom national standard for electrical safety, BS 7671: 2001 *Requirements for Electrical Installations*. The 16th edition is based, as was the 15th edition, on the International Electrotechnical Commission's (IEC) publication 60364, *Electrical installations in buildings*.

▶ **Figure 1.9** The *Electrician's Guide to the Building Regulations*

The structure of BS 7671 is as follows:

Part 1 – Scope, object and fundamental principles
Chapter 11   Scope
Chapter 12   Object and effects
Chapter 13   Fundamental principles

Part 2 – Definitions

Part 3 – Assessment of general characteristics
Chapter 31   Purpose, supplies and structure
Chapter 32   External influences
Chapter 33   Compatibility
Chapter 34   Maintainability

Part 4 – Protection for safety
Chapter 40   General
Chapter 41   Protection against electric shock
Chapter 42   Protection against thermal effect
Chapter 43   Protection against overcurrent
Chapter 44   Protection against overvoltage
Chapter 45   Protection against undervoltage
Chapter 46   Isolation and switching
Chapter 47   Application of protective measures for safety
Chapter 48   Choice of protective measures as a function of external influences

Part 5 – Selection and erection of equipment
Chapter 51   Common rules
Chapter 52   Selection and erection of wiring systems
Chapter 53   Switchgear
Chapter 54   Earthing arrangements and protective conductors
Chapter 55   Other equipment
Chapter 56   Supplies for safety services

Part 6 – Special installations or locations
Section 600   General
Section 601   Locations containing a bath or shower
Section 602   Swimming pools
Section 603   Hot air saunas
Section 604   Construction site installations
Section 605   Agricultural and horticultural premises
Section 606   Restrictive conductive locations
Section 607   High protective conductor currents
Section 608   Caravans, motor caravans and for caravan parks
Section 611   Installation of highway power supplies, street furniture and street located equipment

Part 7 – Inspection and testing
Chapter 71   Initial verification
Chapter 72   Alterations and additions to an installation
Chapter 73   Periodic inspection and testing
Chapter 74   Certification and reporting

Appendices

▶ **Figure 1.10**
BS 7671: 2001
Requirements for Electrical Installations (the IEE Wiring Regulations) including amendments 1 and 2

1.2.14 The *Electricity Safety, Quality and Continuity Regulations 2002, Statutory Instrument 2002 No. 2665*

The *Electricity Safety, Quality and Continuity Regulations* (ESQCR) are aimed at protecting the general public and consumers of electricity from danger. In addition, the Regulations specify power quality and supply continuity requirements to ensure an efficient and economic electricity supply service for consumers.

The regulations point to BS 7671 as the standard for new electrical installations being connected to distribution networks (regulation 25). Before connection, the consumer may have to satisfy the distributor that the installation is safe and technically sound by providing evidence that the installation complies with the requirements of BS 7671.

Where installations are connected to an alternative source of energy (whether switched alternative supply or operating in parallel to the distribution network), the regulations require the installation to comply with BS 7671 (regulations 21 and 22).

The regulations include a general requirement on generators, distributors and meter operators to construct, install, protect (both electrically and mechanically), use and maintain their equipment to prevent danger so far as is reasonably practicable (regulation 3(1)). There is also a requirement for generators and distributors to inspect their networks, and in the case of overhead lines and substations, maintain such records for at least ten years (regulation 5).

The regulations contain a specific requirement for distributors and meter operators to maintain their equipment on consumers' premises, e.g. cut-outs and service cables (regulation 24(1)).

There is also a requirement for generators, distributors and meter operators to report accidents involving members of the public coming into contact with their equipment and near misses to the DTI Secretary of State (regulation 31).

On request, the distributor must provide a written statement of:

▶ the number of phases, frequency and voltage
▶ the type of earthing system applicable to the connection
▶ the type and rating of the distributor's protective device or devices nearest to the supply terminals
▶ the maximum prospective short-circuit current at the supply terminals
▶ for low voltage connections, the maximum earth loop impedance of the earth fault path outside of the installation.

(Regulation 28 refers.)

1 Protective Multiple Earthing

The distributor must make available an earth connection for new supplies, unless inappropriate for reasons of safety (regulation 24(4) refers). In the UK, a PME[1] TN-C-S[2] supply will, in most cases, be provided.

2 A TN-C-S system is an electrical system in which the neutral and protective functions are combined in a single conductor in part of the system.

1.2.15 *Electrical safety on construction sites*

The HSE publication, *Electrical safety on construction sites* gives advice on precautions that can be taken to reduce the risk of accidents during the construction phase. It covers:

▶ the installation and use of the temporary site distribution system (designed to distribute and supply electricity to plant, work equipment, lighting, site offices and site huts etc.)

- the use of electrical equipment, including portable equipment, supplied by the temporary site distribution system
- commissioning and use of the new permanently fixed electrical installation in the building or structure under construction
- use of an existing fixed electrical installation in a building undergoing substantial modifications, together with the equipment connected to it
- existing electrical installations in buildings or structures about to be demolished
- maintenance of the electrical installation and portable equipment, including suggested inspection and test frequencies.

1.2.16 The *Special Waste Regulations 1996, Statutory Instrument 1996 No. 972*

Special waste includes hazardous or toxic wastes displaying properties such as toxic, very toxic, harmful, corrosive, irritant and carcinogenic. The special waste regulations place requirements relating to the control of the transit, import and export of waste (including recyclable materials), the prevention, reduction and elimination of pollution caused by waste and the requirement for an assessment of the impact on the environment of projects likely to have significant effects on the environment.

Possible waste from electrical work is given in Table 1.3.

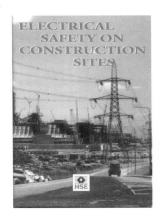

Figure 1.11 HSG 141 (formerly GS 24) *Electrical safety on construction sites*

Table 1.3 Likely waste from electrical work

Description	Waste code
Waste insulating and heat transmission oils and other liquids	1303
insulating or heat transmission oils and other liquids containing PCBs or PCTs	130301
other chlorinated insulating and heat transmission oils and other liquids	130302
non-chlorinated insulating and heat transmission oils and other liquids	130303
synthetic insulating and heat transmission oils and other liquids	130304
mineral insulating and heat transmission oils	130305
Wastes from the electronic industry	1403
chlorofluorocarbons	140301
other halogenated solvents and solvent mixtures	140302
Discarded equipment and shredder residues	1602
transformers and capacitors containing PCBs or PCTs	160201
Batteries and accumulators	1606
lead batteries	160601
Ni-Cd batteries	160602
mercury dry cells	160603
electrolyte from batteries and accumulators	160606
Insulation materials	1706
insulation materials containing asbestos	170601
fluorescent tubes and other mercury containing waste	200201

Hazardous properties of waste include:

- explosive
- oxidising
- highly flammable
- flammable
- irritant
- harmful
- toxic
- carcinogenic
- corrosive
- infectious
- teratogenic[3]
- mutagenic
- ecotoxic
- substances and preparations which release toxic or very toxic gases in contact with water, air or an acid, and
- substances and preparations capable by any means, after disposal, of yielding another substance, e.g. a leachate, which possesses any of the characteristics in this list.

[3] A teratogenic agent is one that affects an embryo or foetus

1.2.17 The *WEEE Directive*

The *Waste Electrical and Electronic Equipment (WEEE) Directive* is Directive 2002/96/EC of the European Parliament and Council and is dated 27 January 2003. The purpose of the directive is, as a first priority, the prevention of waste electrical and electronic equipment and, in addition, the reuse, recycling and other forms of recovery of such wastes so as to reduce the disposal of waste. It also seeks to improve the environmental performance of all operators involved in the life cycle of electrical and electronic equipment, e.g. producers, distributors and consumers and in particular those operators directly involved in the treatment of waste electrical and electronic equipment.

The amount of WEEE generated in the Community is growing rapidly, and the Directive encourages the design and production of electrical and electronic equipment which takes into account and facilitates the dismantling, recovery, reuse and recycling of waste electrical and electronic equipment.

Member states are to ensure systems are set up which include allowing final holders and distributors to return such waste free of charge for suitable treatment and/or recovery.

The Directive applies to:

- large household appliances, including refrigerators, freezers, washing machines, dishwashers, electric stoves, microwaves, electric radiators and air conditioning equipment
- small household appliances, including vacuum cleaners, irons, toasters, grinders, mixers, hair dryers and electric clocks
- IT and telecommunications equipment, including computers, printers, faxes, calculators, telephones, answering machines
- consumer equipment, including radios and televisions, video cameras and recorders, hi-fi systems and musical equipment
- lighting equipment, including luminaires for fluorescent lamps (with the exception of luminaires in households, straight and compact fluorescent lamps, low and high pressure sodium lamps, metal halide lamps)

- electrical and electronic tools (with the exception of large scale stationary industrial tools) including drills, saws, turning, milling, grinding, sawing, cutting and shearing machines, sewing and knitting machines, tools for welding, soldering or similar use, equipment for spraying, spreading and tools for mowing and other garden activities
- toys, leisure and sports equipment including electric train and car sets, video games, computers for biking, driving, running, sports equipment with electric or electronic components, coin slot machines
- medical devices (with the exception of all implanted and infected products) including radiotherapy, cardiology, dialysis, pulmonary ventilators, nuclear medicine, analysers, and appliances for detecting, preventing, monitoring, treating or alleviating illness, injury or disability
- monitoring and control instruments including smoke detectors, heating regulators, thermostats, measuring, weighing or adjusting appliances for the household or laboratories
- automatic dispensers, including automatic dispensers for drinks, bottles, cans, food, solid products and money.

Chapter 12 of this publication gives further information.

1.3 Maintenance strategies

1.3.1 Why perform maintenance?

Maintenance is basically carried out for the following reasons:

- to prevent danger
- to reduce unit cost
- to keep a facility in operation (reliability)
- to prevent pollution of the environment.

To prevent danger

Maintenance for safety may be carried out to meet the following general requirements:

- common law requirements
- explicit legal requirements
- implied legal requirements.

Common law implies a general duty of care to other persons and their livestock, property etc.

Explicit legal requirements are those where there are particular requirements in the legislation for maintenance. This type of maintenance requirement is not common. More frequently, as in the *Electricity at Work Regulations*, the requirement is phrased as follows: 'As may be necessary to prevent danger, all systems shall be maintained so as to prevent, so far as is reasonably practicable, such danger' (regulation 4(2)).

The requirement here is to maintain the system (including equipment) so as to prevent danger, and this may be achieved by carrying out a maintenance activity. In general, equipment cannot be kept in a safe condition without actually being maintained. Normally it is necessary to inspect and/or test a system to determine if maintenance (including repairs) is necessary. It is also likely to be necessary to monitor the effectiveness of any maintenance procedures.

To reduce unit cost and to keep a facility in operation (reliability)
The issues concern minimising business costs and maximising income.

Maintenance carried out to reduce the cost of an enterprise would include action taken to reduce or avoid:

a the cost of failure of plant or equipment – repair costs
b the cost of loss of production – revenue costs
c the cost of loss of service – revenue and goodwill.

▸ *Cost of failure of plant or equipment – repair costs.* Decisions on the approach to be taken are rarely simple. For example, all the lamps of a street lighting installation will need to be replaced at some time or the street will end up in darkness. The decision as to whether it is cost effective to replace the lamps routinely (preventive maintenance) or when they fail (breakdown) will need to take into account the reduction in light output (if significant), the effect on traffic of frequent disturbances and the cost of attending to replace lamps. It might be decided that breakdown maintenance was appropriate for a B road, while preventive lamp replacement was necessary for a motorway.

A balance needs to be achieved between the cost of the maintenance activity and the cost of the equipment.

The cost of maintaining a large motor might well be small compared with the cost of replacement, whereas the cost of replacing a single tungsten filament lamp will far outweigh the cost of the lamp. The lamp may be maintained on a breakdown basis, the motor on a routine basis.

▸ *Loss of production – revenue costs.* In many situations the cost of the failed piece of equipment is insignificant compared with the cost of loss of output or production. In these circumstances breakdown maintenance is unlikely to be appropriate.

▸ *Loss of service – revenue and goodwill.* Customer goodwill is difficult to estimate financially, but should be considered, when determining maintenance regimes. The additional costs of early replacement, or even frequent maintenance, can be justified by customer goodwill.

To prevent pollution of the environment
Maintenance may be required to be carried out, not simply to protect people's health and safety, but also to protect the environment. This may not be cost effective, but it may be seen as a general duty of care or it may be a legislative requirement as required by the *WEEE Directive*, the *Clean Air Act* or the *Environmental Protection Act* and associated regulations.

1.3.2 Maintenance categories
There are three general categories of maintenance:

▸ breakdown
▸ preventive
▸ condition monitored.

Breakdown maintenance

Breakdown maintenance is the simplest approach. Equipment and systems are repaired or replaced when they cease to work. This approach can be followed where breakdown or failure to work does not result in danger, and the consequences of the failure or breakdown are otherwise acceptable. This approach would be reasonable and is generally adopted for the failure of lamps in domestic premises.

Preventive maintenance

Preventive maintenance is carried out before breakdown or failure occurs. This will normally mean that maintenance is planned and carried out at specified intervals, although it may be initiated by a signal or feedback of some description. Examples of preventive maintenance are the routine changing of street lighting lamps and the changing of the engine oil of a generating set.

Condition monitored maintenance

Condition monitored maintenance requires the measurement of certain parameters of the equipment, such as vibration or temperature. At pre-set levels, alarms are initiated allowing the equipment to be shut down and maintenance carried out. This is appropriate for such equipment as large motors where the cost of breakdown is high or where shutdown is to be minimised.

1.3.3 Inspection and testing

Rarely in legislation is there a specific requirement for inspection and testing. The requirement is for maintenance, either at specified intervals or to ensure safety, as in the *Electricity at Work Regulations*. The purpose of inspection and testing is to determine if any maintenance, including repairs, is required. The keeping of inspection and test results will also demonstrate that, as far as the inspector and tester could reasonably determine, maintenance was or was not required at the time of the inspection and test. The records will also enable slow deterioration or step changes in the condition of equipment to be identified.

1.3.4 Keeping of records

Much legislation simply requires either that maintenance be carried out or, more commonly, as in the *Electricity at Work Regulations*, that equipment be maintained so as to prevent, so far as is reasonably practicable, danger. There is generally no specific requirement to keep records. However, in most of the guidance issued by the Health and Safety Executive, including the *Memorandum of Guidance on the Electricity at Work Regulations 1989*, the advice is given that records of maintenance including test results, preferably kept throughout the working life of an electrical system, will enable the condition of the equipment and the effectiveness of the maintenance policies to be monitored. Additionally, without effective monitoring, duty holders may find it difficult to demonstrate that the requirement for maintenance has been met.

Risk assessment, manual handling, display screen regulations, safety signs and signals

2

2.1 Risk assessment

2.1.1 Legislation

The *Management of Health and Safety at Work Regulations* (MHSWR) impose duties with respect to the assessment of risks upon every employer and self-employed person.

Coverage of the regulations is given in Table 2.1.

Regulation	Application
1	Citation, commencement and interpretation
2	Disapplication of these regulations
3	Risk assessment
4	Principles of prevention to be applied
5	Health and safety arrangements
6	Health surveillance
7	Health and safety assistance
8	Procedures for serious and imminent danger and for danger areas
9	Contact with external services
10	Information for employees
11	Cooperation and coordination
12	Persons working in host employers' or self-employed persons' undertakings
13	Capabilities and training
14	Employees' duties
15	Temporary workers

▶ **Figure 2.1**
Management of Health and Safety at Work Regulations 1999, Approved Code of Practice L21

▶ **Table 2.1** Coverage of the MHSWR

continues

© The Institution of Engineering and Technology

▶ Table 2.1 *continued*

Regulation	Application	
16	Risk assessment in respect of new or expectant mothers	
17	Modification of the Safety Representatives and Safety Committee Regulations 1977	
18	Notification by new or expectant mothers	
19	Protection of young persons	
20	Exemption certificates	
21	Provision as to liability	
22	Exclusion of civil liability	
23	Extension outside Great Britain	
24, 25, 26, 27	Amendments to various regulations	
28	Regulations to have effect as health and safety regulations	
29	Revocations and consequential amendments	
30	Transitional provision	

Regulation 3 – Risk assessment requires that every employer and self-employed person must carry out a suitable and sufficient risk assessment of the risks to the health and safety of his employees while they are at work and the risks to the health and safety of persons not in his employment arising out of or in connection with the conduct by him of his undertaking. Relevant statutory provisions must be considered. Any assessment made must be reviewed by the employer or self-employed person if it may no longer be valid or if there has been a significant change in the circumstances. An employer must not employ a young person unless a risk assessment has been made which must take particular account of factors including:

▶ that person's inexperience, lack of awareness and immaturity
▶ the fitting out and layout of the workplace and the workstation
▶ the nature, degree and duration of exposure to physical, biological and chemical agents
▶ the form, range and use of work equipment and the way in which it is handled
▶ the organisation of processes and activities
▶ the extent of the health and safety training provided or to be provided to young persons, and
▶ the risks from agents, processes and work.

Where the employer employs five or more persons, he must record the significant findings of the assessment and any group of his employees identified by it as being especially at risk.

Regulation 4 – Health and safety arrangements. Where an employer implements any preventive and protective measures, this must be done on the basis of the principles specified in Schedule 1 to the MHSWR given in Table 2.2.

Schedule 1 to the MHSWR: general principles of prevention
Avoid risks
Evaluate the risks which cannot be avoided
Combat the risks at source
Adapt the work to the individual, especially as regards the design of workplaces, the choice of work equipment and the choice of working and production methods, with a view, in particular, to alleviating monotonous work and work at a predetermined work rate and to reducing their effect on health
Adapt to technical progress
Replace the dangerous by the non-dangerous or the less dangerous
Develop a coherent overall prevention policy which covers technology, organisation of work, working conditions, social relationships and the influence of factors relating to the working environment
Give collective protective measures priority over individual protective measures
Give appropriate instructions to employees

▶ **Table 2.2** Schedule 1 to the MHSWR: general principles of prevention

Regulation 5 – Health and safety arrangements requires every employer to make arrangements for the effective planning, organisation, control, monitoring and review of the preventive and protective measures. Once again, where an employer employs five or more employees, the arrangements must be recorded.

Regulation 6 – Health surveillance requires that employees are provided with health surveillance where there are risks to their health and safety. These procedures may be necessary where there are risks of exposure to chemicals, radiation, extreme temperatures, infections etc.

Regulation 7 – Health and safety assistance requires the appointment of one or more competent persons to assist in complying with the regulations. This may mean that in particular instances specialists will have to be specifically employed to advise on the risks and the action to be taken.

Regulation 8 – Procedures for serious and imminent danger and for danger areas. Where there are particular risks presenting serious and imminent danger such as fire, explosion, radiation etc. procedures must be established, a sufficient number of competent persons must be nominated and no employees must have access to any areas where it is necessary to restrict access on health and safety grounds unless the employee concerned has received adequate health and safety instruction. There must be set procedures for potentially dangerous situations, e.g. fire alarm, bomb alert, and all employees should know these as part of their induction training.

Regulation 9 – Contacts with external services. Employers must ensure that any necessary contacts with external services, such as first-aid, emergency medical care and rescue work, are arranged.

Regulation 10 – Information for employees. All employees must be provided with basic health and safety information which must be comprehensible and take into account any disabilities or lack of English as a first language. Any employer, before employing a child, shall provide the parent/guardian with comprehensive and relevant health and safety information including any risks and protective measures.

Regulation 11 – Cooperation and coordination. If more than one employer is using a site, e.g. university and NHS Trust, they should coordinate their health and safety obligations as far as possible and inform each other of potential risks and hazards. The appointment of a Health and Safety coordinator is the best solution if many temporary or short-term staff are employed.

Regulation 12 – Persons working in host employers' or self-employed persons' undertakings. External contractors (e.g. service engineers) should be informed of any risks pertaining to the area they will be in, and also what the emergency evacuation procedures are.

Regulation 13 – Capabilities and training. Employees should receive the necessary training at each change in working practice or introduction of new technique. Each person should always be capable of working within their limitations. Particular attention should be paid to young persons and new recruits. Regular refresher courses are recommended. All training should be during working hours or treated as an extension of working time.

Regulation 14 – Employees' duties. Employees have a duty to take reasonable care for their own Health and Safety and also for others who may be affected by their actions. They must inform the employer of any dangerous situation or shortcomings not covered by the risk assessments in place. Any accidents or incidents must be reported.

Regulation 15 – Temporary workers. Fixed-contract employees must be supplied with any relevant information on safety and health surveillance before they start their duties.

Regulations 16, 17, 18 – Risk assessment in respect of new or expectant mothers. Employers may need to alter working conditions or hours of work to avoid risks of exposure to certain chemicals etc. If these risks cannot be avoided then the woman can be suspended from work until the risk is past. Employees involved in night working can be suspended from this on production of a medical certificate. Employees must notify their employers in writing and produce a certificate of pregnancy.

Regulation 19 – Protection of young persons. Young persons must be supervised in training and have their risks reduced to the lowest level possible.

2.1.2 What is a risk assessment?

A risk assessment is nothing more than a careful examination of what, in your work, could cause harm to people, so that you can weigh up whether enough precautions have been taken or more should be done to prevent harm. The aim is to ensure no one gets hurt or becomes ill.

The important thing is to decide whether a hazard is significant and whether it is covered by satisfactory precautions so that the risk is small.

Hazard Means anything that can cause harm (e.g. electric shock, high temperatures, moving mechanical parts, electric testing, driving, slips, trips, falls, use of ladders and step-ladders).

Risk Is the likelihood, high or low, that the hazard will cause a specified harm to someone or something.

For instance, electricity can kill (presents a high hazard) but the risk of it doing so in an office environment is remote (risk is small), provided that equipment is properly selected, used and maintained.

It is important that the persons carrying out risk assessments are competent to do the work. Individuals who consider themselves not competent to carry out the risk assessments entrusted to them must advise their employer.

Risk assessments are not difficult and need not be complicated. In most firms in the commercial and light industrial sectors, the hazards are few and simple.

The risk assessment must be carried out before the work that gives rise to the risk is performed.

There are five steps to risk assessment:

1 Look for the hazards
2 Decide who might be harmed and how
3 Evaluate the risks and decide whether the existing precautions are adequate or whether more should be done
4 Record the findings
5 Review the assessment and revise if necessary.

Look for the hazards

Walk around the workplace and look at what could reasonably be expected to cause serious harm or affect several people. Employees should be consulted. Manufacturer's instructions/data sheets can help to spot hazards, as can experience and accident and ill-health records.

Decide who might be harmed and how

Do not forget:

▶ that young workers, trainees, new workers and expectant mothers may be particularly at risk
▶ to consider cleaners, visitors, contractors, maintenance workers and part time workers who may not be in the workplace all the time
▶ members of the public or people from other organisations if there is any possibility they could be hurt by the activities.

Evaluate the risks and decide whether existing precautions are adequate or more should be done

Consider how likely it is that each hazard could cause harm. This will determine whether or not something needs to be done to reduce the risk. Even after all precautions have been taken, some risk usually remains. For each significant hazard it has to be decided whether this remaining risk is high, medium or low.

First, have all the things been done that the law requires? For example, the *Electricity at Work Regulations 1989* give the requirements shown in Table 2.3.

▶ **Figure 2.2** Guarded hole in floor

If a requirement in a regulation is absolute, for example it is not qualified by the words 'so far as is reasonably practicable', the requirement must be met regardless of cost or any other consideration. Certain of the regulations making such absolute requirements are subject to the defence provision of regulation 29.

Table 2.3 Requirements placed by the *Electricity at Work Regulations 1989* (refer to regulations for full details)

Requirements	Regulation No.
Persons on whom duties are imposed	3
The work must not give rise to danger	4(3)
Protective equipment must be suitable and properly maintained	4(4)
Earthing or other suitable precautions may need to be taken	8
Integrity of referenced conductors must not be compromised	9
Equipment made dead must not be allowed to become live if danger could arise	13
Work must not be performed on live conductors or equipment, unless **a** it is unreasonable in all the circumstances for the conductor or equipment to be made dead **b** it is reasonable in all the circumstances for work to be performed on live conductors or equipment **c** suitable precautions (including, where necessary, the provision of suitable protective equipment) are taken to prevent injury	14
Working space, access and lighting	15
Persons to be competent to prevent danger and injury	16

Someone who is required to do something 'so far as is reasonably practicable' must assess, on the one hand, the magnitude of the risks of a particular work activity or environment, and, on the other hand, the costs in terms of physical difficulty, time and trouble and expense which would be involved in taking steps to minimise or eliminate those risks. If, for example, the risks to health and safety of a particular work process are very low and the costs or technical difficulties of taking certain steps to prevent those risks are very high, it might not be reasonably practicable to take those steps. The greater the degree of risk, the less weight that can be given to the cost of measures needed to prevent that risk.

In the context of the *Electricity at Work Regulations*, where the risk is very often that of death, for example from electrocution, and where the nature of the precautions that can be taken is so very often cheap and simple (e.g. isolation), the level of duty to prevent that danger approaches that of an absolute duty.

There are other legal requirements including:

▷ *Management of Health and Safety at Work Regulations 1999* (Management Regulations)
▷ *Manual Handling Operations Regulations 1992* (as amended by the *Health and Safety (Miscellaneous Amendments) Regulations 2002*) (Manual Handling Regulations)
▷ *Personal Protective Equipment at Work Regulations 1992* (PPE)
▷ *Health and Safety (Display Screen Equipment) Regulations 1992* (as amended by the *Health and Safety (Miscellaneous Amendments) Regulations 2002*) (Display Screen Regulations)
▷ *Noise at Work Regulations 1989* (Noise Regulations)

▶ *Control of Substances Hazardous to Health Regulations 2002* (COSHH)
▶ *Control of Asbestos at Work Regulations 2002* (Asbestos Regulations)
▶ *Control of Lead at Work Regulations 2002* (Lead Regulations).

Generally accepted industry standards should be in place, such as the provision of suitable and safe test equipment and protective equipment such as safety boots.

Questions to be asked when performing an electrical risk assessment include:
Can the work be done with the equipment dead or energised at a safe voltage or current?
Is it absolutely necessary in all the circumstances for someone to be working on or near to equipment that is live at dangerous voltages or current levels?
What is the maximum voltage on conductors that will be exposed during the work activity?
Are the electricians competent? Are they adequately trained and knowledgeable to perform the work without putting themselves or others at risk?
If electricians are not fully competent, are they adequately supervised?
Has the circuit to be worked on been correctly identified?
Is a permit to work procedure needed?
Will the isolation procedures prevent injury?
▶ are the correct circuits being isolated? is there any possibility of a back feed?
▶ are the isolation procedures secure? is a padlock to be used?
▶ have the **circuits** been proved dead?
▶ have all **circuit conductors** been proved dead, including protective conductors (in case of a wiring fault)?
▶ is a proving unit available? has the test instrument been proved before and after use?
Is there any possibility of a borrowed neutral situation?
Have notices been posted?
Can physical safeguards be applied to the equipment to prevent injury?
Is the test instrumentation of a safe design and has it been properly maintained?
Is access available in the form of scaffolding or will a step-ladder be needed?

▶ **Table 2.4** Electrical risk assessments

In addition to the list of industry accepted standards, the law also says that the workplace must be kept safe, as far as is reasonably practicable. The real aim is to make all risks small by adding to the list of precautions as necessary. If something needs to be done, an 'action list' should be drawn up and priority given to any remaining risks that are high and/or those which could affect most people. In taking action the following must be considered:

a Is it possible to eliminate the hazard?
b If not, how can the risks be controlled so that harm is unlikely?
c In controlling risks the principles below should be applied, if possible, and in the following order:
  ▶ is there a lower risk option?
  ▶ prevent access to the hazard (e.g. by guarding)
  ▶ organise work to reduce exposure to the hazard
  ▶ issue personal protective equipment.
d Provide welfare facilities (e.g. washing facilities for removal of contamination and first-aid facilities).

Improving health and safety need not cost a lot. For instance, wearing safety shoes, using test probes that meet the recommendations of GS38, fitting a hands-free phone in the van, or putting some non-slip material on slippery steps, are inexpensive precautions when considering the risks. And failure to take simple precautions can cost a lot more if an accident does happen.

For many electricians, the work varies and they often move from one site to another. The hazards that can reasonably be expected should be identified and the risks assessed. After that, the site should be checked for additional hazards, including getting information from others on site, and taking whatever action is necessary.

If a workplace is shared by other employees and self-employed people, they should be informed of any risks that the work could cause them and the precautions that are being taken. For example, during electrical work, it is very likely the electrician will need to isolate part or all of the installation, and hence power to that part of the premises will not be available for the duration of the work.

Record the findings

If fewer than five employees are involved, it is not necessary to write anything down, although it is useful to keep a written record of what has been done. If five or more people are involved, the significant findings of the assessment must be recorded. This means writing down the significant hazards and conclusions.

Examples might be:

Electrical installation: insulation and earthing checked and found sound or
Fume from welding: local exhaust ventilation provided and regularly checked

Employees must be informed about the findings.

Suitable and sufficient – not perfect! Risk assessments must be suitable and sufficient. It is sufficient to show that:

▶ a proper check was made
▶ you asked who might be affected
▶ you dealt with all the obvious significant hazards, taking into account the number of people who could be involved
▶ the precautions are reasonable, and the remaining risk is low.

The written record should be kept for future reference or use. It can help if an inspector asks what precautions were taken, or if there is any action for civil liability. It can also serve as an aide-mémoire for particular hazards and precautions. And it helps to show that the requirements of the law have been met.

Review the assessment and revise if necessary

Changes in conditions, such as cable colour changes, new instruments and the employment of new staff (for example, an apprentice or summer trainee), could lead to the need to review the risk assessment. If there is any significant change, the assessment must be added to, taking account of the new hazard. There is no need to amend the assessment for every trivial change, or even for each new job. However, if a new job introduces significant new hazards of its own, then these should be considered in their own right and appropriate action taken to keep the risks down.

In any case, it is good practice to review the risk assessment from time to time to make sure that the precautions are still working effectively.

▶ **Table 2.5** Guide to performing a risk assessment

Step 1 The hazard	Step 2 Who might be harmed?	Step 3 Is more control needed to control the risk?	Step 4	Step 5 Review and revision
Possible hazards include: Chemicals such as battery acids Electric burns Electric shock Ladders Low temperatures Manual handling Moving parts of machinery such as fan blades Noise Outdoor conditions Overheating of electrical equipment Particles from drilling, grinding or cutting Poor access Poor lighting Pressure systems Risk of fire, overheating Scaffolding Slipping Tripping Vehicles, including fork lift trucks Working at height	Groups of people that could be affected: Staff Other contractors Operators Cleaners And, in particular: Persons with disabilities Inexperienced personnel Visitors Lone workers	For the hazards listed, do the precautions taken Meet the legal requirements? Comply with a recognised industry standard? Represent good practice? Reduce risk as far as is reasonably practicable? Have the following been provided: Adequate information, instruction or training? Adequate systems or procedures?	Record the findings	Set a date for a review of the assessment Check the precautions for each hazard adequately control the risk Making changes in the workplace such as new machines, substances and procedures may require a risk assessment review.

A risk assessment form is shown in Figure 2.3.

Table 2.6 Guide to control measures

Control measures could include:
Hard hat, Hi-Vis (High Visibility) vest and steel toe-capped boots worn when on site
Floorboards, in the immediate area, will be securely fixed before any work takes place from step-ladders. This work will be the responsibility of the main contractor and work will not proceed until the boards are secured
Barriers around openings in floors or ceilings to be provided
Provision of suitable and safe ladders, scaffolding or other access equipment
Step-ladders to be visually inspected before use
Any defective ladders to be reported to the Supervisor and taken out of use
Step-ladders must be used on a level, firm surface
One operative to work from the step-ladder at any one time
Operative to have three points of bodily contact at all times while using step-ladders
Step-ladders to be used in accordance with manufacturer's recommendations
All tools to be carried in a shoulder bag or holster when using step-ladders
All personal clothing to be tight and secure with no loops or loosely hanging straps
Isolation certificate to be issued when DB3 is isolated and locked off. Certificate and keys to be held by site supervisor
Only 110 V electrical equipment to be used and all portable electrical equipment to be within PAT testing period and bear a date tag
All portable electrical equipment to be visually inspected on a daily basis
Extension leads to be routed overhead, away from walkways, from transformer to point of use
Extension leads must not be used in wet areas or conditions
Care to be taken while other works are in operation
Carry out regular noise assessments, record findings
Operatives to wear gloves when handling metalwork and pulling cables
Goggles, ear plugs and gloves to be worn when drilling, cutting and grinding
Safe access and safe exit; in both normal and emergency conditions
Special precautions if there are restricted workspaces
Provision of adequate lighting, both normal and emergency
Precautions to be taken for lifting and handling
Fire precautions to be implemented if required
Mechanical handling devices
Precautions to be taken if brazing and welding equipment, including flammable gases e.g. acetylene, dangerous gases e.g. oxygen, and asphyxiating gases e.g. argon to be used
Good housekeeping at all times

Risk assessment form			Assessment by	
Company name			(Name and signature)	
Address			Date of assessment	
			Recommended date of review	
Project number:			(step 5)	
The hazard (List the hazards)	**Those at risk** (List the persons in groups at risk)	**Evaluation of controls** (List existing controls and procedures)	**Findings** (List risks not adequately controlled)	**Implementation of findings** Record action taken Date Action
(step 1)	(step 2)	(step 3)	(step 4)	

▶ **Figure 2.3** Risk assessment form

2.1.3 Numeric risk assessment

Numeric risk assessments may need to be performed for certain projects.

A figure between 1 and 4 should be assigned to the severity of the injury, where 1 represents a minor injury (such as a cut) and 4 represents a major injury (such as electric shock). In addition, built into these four levels is the number of people that could be injured. A minor injury to one person would be a '1' whereas a minor injury to three or four people might increase the rating to a '2'. The figure assigned represents the seriousness of the injury and the number of people involved before any preventive measures are implemented.

Next, a figure should be entered for the likelihood of the injury occurring, i.e. the risk. A '1' represents something that is relatively unlikely to occur whereas a '4' represents something that occurs frequently, for example sharp particles coming from an angle grinder when in use.

The two figures are then multiplied together to produce the risk rating. The figure is compared with the matrix (Figure 2.4). From the matrix, a figure of 4 or below means a serious accident is not likely to happen. A figure of 6 to 9 means that an accident may happen and preventive measures should be considered. A figure of 12 or above means that a serious accident is likely to happen unless preventive measures are taken.

The risk rating can be determined from the matrix in Figure 2.4.

Next, for risk ratings of 6 and above, suitable preventive measures must be considered and written down. The preventive measure should reduce the likelihood of injury. A new risk rating is then given for the particular hazard and the process repeated, if necessary, until the risk rating is reduced to a value of 4 or less.

For example, consider a job where it is necessary to remove redundant wiring from a trunking system which contains some live cables. There is a real risk of electrocution. The severity of the injury is major (i.e. '4') and, without preventive measures, is likely to occur (i.e. '3'). This gives a risk rating of 12. A preventive measure of isolation for the particular circuits must be implemented. The risk rating now needs to be reconsidered. The severity of the injury is still a '4' but the likelihood of the injury has been reduced. Since there is still a possibility of electrocution due to other cables in the trunking being energised, there is still a possibility of electrocution so the risk rating is now a '4'

Figure 2.4 Risk rating matrix

Severity Of Injury	Likelihood of Injury			
	Frequent 4	Likely 3	Possible 2	Improbable 1
Major 4	High 16	High 12	Medium 8	Low 4
Severe 3	High 12	Medium 9	Medium 6	Low 3
Serious 2	Medium 8	Medium 6	Low 4	Low 2
Minor 1	Low 4	Low 3	Low 2	Low 1

multiplied by a '2' (possible). This is still not acceptable, so a further preventive measure is needed such as the isolation of all circuits in the trunking. This would reduce the possibility of electrocution to 'improbable' with an acceptable risk rating of 4.

2.1.4 Live working
Regulation 14 of the *Electricity at Work Regulations 1989* states:

'No person shall be engaged in any work activity on or so near any live conductor (other than one suitably covered with insulating material so as to prevent danger) that danger may arise unless

a it is unreasonable in all the circumstances for it to be dead; and
b it is reasonable in all the circumstances for him to be at work on or near it while it is live; and
c suitable precautions (including where necessary the provision of suitable protective equipment) are taken to prevent injury.'

The *Memorandum of Guidance to the Electricity at Work Regulations 1989* advises that work should be carried out live only when there is no other way of reasonably effecting the work. Working includes testing and fault finding. Procedures for working on or near live parts, if allowed at all, must be included in the safety instructions issued to staff. Where the work is of a repetitive nature, such as the routine testing of a particular piece of equipment, it would be reasonable to expect specific test regimes or test equipment to be fabricated that would allow the test and measurements to be carried out without danger to the staff involved.

2.2 Manual handling

The *Manual Handling Operations Regulations 1992* came into force on 1 January 1993. They were made under the *Health and Safety at Work etc. Act 1974* and implemented *European Directive 90/269/EEC*. The Health and Safety Executive publish guidance on these regulations called *Manual Handling* and the reference number is L23.

▶ **Figure 2.5** L23. *Guidance on the Manual Handling Regulations*

Manual handling means any transporting or supporting of a load (including the lifting, putting down, pushing, pulling, carrying or moving thereof) by hand or bodily force. The Regulations apply to the manual handling of loads, i.e. by human effort, as opposed to mechanical handling by crane or fork-lift truck. Introducing mechanical assistance, for example a sack truck or wheelbarrow, may reduce but not eliminate mechanical handling since human effort is still required to move, steady or position the load.

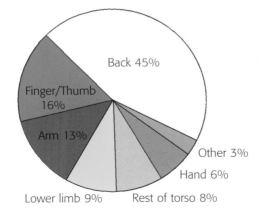

▶ **Figure 2.6** Injuries caused by manual handling

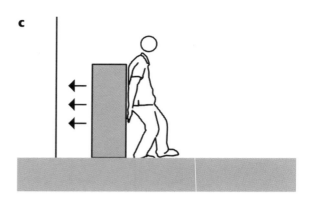

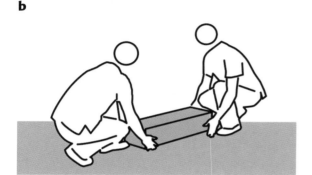

▶ **Figure 2.7** Manual handling

a A simple sack trolley can significantly reduce the amount of manual handling
b Team handling
c Using the strong leg muscles
d A hoist can be used

The Regulations require that each employer must:

a so far as is reasonably practicable, avoid the need for employees to undertake any manual handling operations at work which involve the risk of their being injured
b where it is not reasonably practicable to avoid the need for employees to undertake any manual handling operations at work which involve the risk of their being injured
  i make a suitable and sufficient assessment of all such manual handling operations to be undertaken by employees having regard to the factors as specified in column 1 of schedule 1 of these regulations and considering the questions which are specified in the corresponding entry in column 2 of that schedule
  ii take appropriate steps to reduce to a reasonable minimum the risk of injury to those employees arising out of their undertaking any such manual handling operations
  iii take appropriate steps to provide any of those employees who are undertaking such manual handling operations with general indications and, where it is reasonably practicable to do so, precise information on:
    ▶ the weight of each load
    ▶ the heaviest side of any load whose centre of gravity is not positioned centrally.

(Regulation 4(1) refers.)

The factors to which an employer must have regard and the questions which he must consider in making an assessment are reproduced in Figure 2.8 which is given in schedule 1 of the *Manual Handling Operations Regulations*.

Manual handling of loads
Manual handling assessment checklist

Client _____

Address _____

_____ Postcode _____

Summary of assessment	Overall priority for remedial action
Operations covered by this assessment	Nil/Low/Medium/High
_____	Remedial action to be taken
_____	
_____	_____
Locations _____	Date for reassessment _____
Personnel involved _____	Assessor's name _____
Date of assessment _____	Signature _____

Section A – Preliminary

1 Do the operations involve a significant risk of injury. If yes go to Q2. If no the assessment need go no further.   yes/no

2 Can the operations be avoided/mechanised/automated at reasonable cost   yes/no
If no go to Q3. If yes proceed and then check results satisfactory.

3 Are the operations clearly within the guidelines attached. If no go to Section B (overleaf).   yes/no
If yes, go straight to Section C.

Section B – More detailed assessment, where necessary   (see second page)

Section C – Overall assessment of risk

4 What is your overall assessment of the risk of injury?

insignificant/low/medium/high

Section D – Remedial action

5 What remedial steps should be taken in order of priority?

i _____

ii _____

iii _____

iv _____

v _____

And finally   Complete the summary above
Compare it with your other manual handling assessments
Decide your priorities for action
Take the action and check that it has the desired effect

Page 1

▶ **Figure 2.8a** Form for risk assessment

Section B – More detailed assessment where necessary						
Questions to consider (If the answer to a question is yes, place a tick against it and then consider the level of risk.)		Level of risk (tick as appropriate)			Possible remedial action
	yes	low	medium	high	
The task – does it involve					
▶ holding or manipulating a load at a distance from the trunk of the body?					
▶ twisting the trunk?					
▶ stooping?					
▶ reaching upwards?					
▶ large vertical movement?					
▶ long carrying distances?					
▶ strenuous pushing or pulling?					
▶ unpredictable movement of the load?					
▶ repetitive handling?					
▶ insufficient rest or recovery?					
▶ a rate of work imposed by a process?					
The load – is it					
▶ heavy?					
▶ bulky or unwieldy?					
▶ difficult to grasp?					
▶ unstable, or with contents likely to shift?					
▶ sharp, hot or otherwise potentially damaging?					
The working environment – are there					
▶ space constraints preventing good posture?					
▶ uneven, slippery or unstable floors?					
▶ variations in level of floors or work surfaces?					
▶ extremes of temperature or humidity?					
▶ conditions causing ventilation problems or gusts of wind?					
▶ poor lighting conditions?					
Individual capability – does the job					
▶ require unusual strength, height, etc.?					
▶ create a hazard to those who might reasonably be considered to be pregnant or to have a health problem?					
▶ require special information or training for its safe performance?					
Other factors					
Is movement or posture hindered by personal protective equipment or by clothing?					Page 2

▶ **Figure 2.8b** Form for risk assessment

Electrical Maintenance

© The Institution of Engineering and Technology

> **Figure 2.9** Beginning to lift the load

Basic lifting

Stop and think. Plan the lift. How heavy is the load? Rocking it may give an indication. Where does the load have to go? Do I need handling aids? Do I need help? Remove obstructions such as wrapping. Can I reduce the weight of the load? For a long lift, such as from floor to shoulder height, will it help to rest the load midway such as on a table in order to take a rest and to change grip.

Place the feet. Feet apart, giving a stable and balanced base for lifting. Unsuitable footwear or tight clothing will make the lifting more difficult.

Adopt a good posture. Bend the knees so that the hands, when grasping the load, are as nearly level with the waist as possible, but do not kneel or overflex the knees. Keep the back straight. Lean forward slightly over the load to get a good grip. Keep shoulders level and facing in the same direction as the hips.

Get a firm grip. Try to keep the arms within the boundary formed by the legs. Different people may prefer a different grip, but it must be secure. A hook grip is less fatiguing than keeping the fingers straight. If it is necessary to adjust the grip while the lift proceeds, do this as smoothly as possible.

Don't jerk. Carry out the lifting as smoothly as possible, keeping control of the load.

Move the feet. Do not twist the trunk when turning to the side.

> **Figure 2.10** Correct lifting

Keep close to the load. Keep the load as close to the trunk for as long as possible. Keep the heaviest side of the load next to the trunk.

Put the load down, and then adjust it. Put the load down and then slide it into position as necessary.

2.3 Display screen equipment

The *Health and Safety (Display Screen Equipment) Regulations* require a risk assessment to be carried out by the employer at each visual display unit workstation used in the purpose of the business or provided by him and used by operators. Guidance on the regulations is provided by the Health and Safety Executive in its publication L26 *Work with display screen equipment*. Any person responsible for assessment will need to obtain a copy of HSE document L26 which includes, in Appendix 5, a workstation checklist which can be used as an aid to risk assessment and to help comply with the Schedule to the Regulations.

Those responsible for assessments must be familiar with the requirements of the regulations and be able to:

▶ assess the risks associated with the workstation
▶ make a clear record of the assessment and communicate the findings to those who need to take action
▶ assess any action necessary
▶ recognise their own limitations and, as necessary, call upon further expertise.

It is necessary to review the assessments from time to time, particularly if there has been:

▶ a major change in workstation furniture
▶ a major change to the hardware
▶ a major change to the software
▶ changes in task requirements.

It will also be necessary to review an assessment if the workstation is removed or there are significant changes in the environment such as in the level of lighting, either natural or artificial.

Items to consider, as detailed in the schedule to the regulations, include:

▶ adequate lighting
▶ adequate contrast, no glare or distracting reflections
▶ distracting noise minimised
▶ leg room and clearances to allow postural changes
▶ window covering, if needed, to minimise glare
▶ software: appropriate to the task, adapted to the user, providing feedback on the system status, no undisclosed monitoring
▶ screen: stable image, adjustable, readable, glare/reflection free
▶ keyboard: usable, adjustable, detachable, legible
▶ work surface: with space for flexible arrangement of equipment and documents; glare free
▶ chair: adjustable and stable
▶ footrest, if needed.

▶ **Figure 2.11** L26. *Work with display screen equipment*

▶ **Figure 2.12** Subjects covered in the schedule to the *Health and Safety (Display Screen Equipment) Regulations*

2.4 Safety signs and signals

2.4.1 Introduction
The *Health and Safety (Safety Signs and Signals) Regulations 1996* implemented *European Council Directive 92/58/EEC* setting minimum requirements for the provision of safety signs at work. Following the risk assessment carried out as required by the *Management of Health and Safety at Work Regulations* all necessary safety signs and signalling procedures must comply with the *Health and Safety (Safety Signs and Signals) Regulations*. The Health and Safety Executive publish guidance on these regulations, *Safety signs and signals* (publication L64).

The Regulations require employers to ensure that safety signs are provided (or are in place) and maintained in circumstances where risks to health and safety have not been avoided by other means, e.g. engineering controls or safe systems of work.

In determining where to provide safety signs, employers need to take into account the risk assessment made under the MHSWR 1999 (*Management Regulations*). This assessment deals with hazard identification, the risks associated with those hazards and the control measures to be taken. When the control measures identified in the risk assessment have been taken, there may be a 'residual' risk, such that employees need to be warned, and informed of any further measures necessary. Safety signs are needed if they will reduce this residual risk. If the risk is not significant, there is no need to provide a sign.

Emergency escape and first-aid signs are shown, as are fire fighting signs.

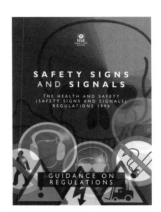

▸ **Figure 2.13** *Safety signs and signals*, publication L64 from the HSE

2.4.2 Sign colours
The Regulations require specific colours for particular signs. Table 2.7 is reproduced from Schedule 1 of the Regulations.

Colour	Meaning or purpose	Instructions and information
Red	prohibition sign danger alarm	dangerous behaviour, stop, shut-down, emergency cut-out devices, evacuate
	fire fighting equipment	identification and location
Yellow or amber	warning sign	be careful, take precautions, examine
Blue	mandatory sign	specific behaviour or action, e.g. wear personal protective equipment
Green	emergency escape first-aid sign	doors, exits, escape routes, equipment and facilities
	No danger	Return to normal

▸ **Table 2.7** Colours of safety signs

These colour codes are shown pictorially for prohibition, warning, mandatory and emergency escape signs in Figure 2.14.

a

b

c

d

▶ **Figure 2.14**
  a prohibition sign red/white/black – a sign prohibiting behaviour likely to increase or cause danger (e.g. no smoking);
  b warning sign yellow/black – a sign giving warning of a hazard or danger (e.g. danger: electricity);
  c mandatory sign blue/white – a sign prescribing specific behaviour (e.g. eye protection must be worn);
  d emergency escape sign or first-aid signs – a sign giving information on escape routes and emergency exits, first-aid or rescue facilities.

a

Guidance on the format for a wider range of signs is given in HSE document L64.

2.4.3 Signs
Fire fighting signs

Signs are specified for fire fighting equipment and they should be of rectangular or square shape and have white pictograms on a red background (the red part to take up at least 50 per cent of the area of the sign). Examples are reproduced in Figure 2.15.

b

Emergency escape or first-aid signs

Signs are specified for emergency escape or first-aid signs and they should be of rectangular or square shape and have white pictograms on a green background (the green part to take up at least 50 per cent of the area of the sign). Examples are reproduced in Figure 2.16.

c

Signs on containers and pipes

Containers, tanks and vessels used in the workplace to contain dangerous substances, and visible pipes in the workplace containing or transporting dangerous substances will, in general, need to have signs or labels fixed to them unless the risk is adequately controlled or is not significant. There are permitted exceptions. It may not be necessary to affix signs to:

▶ a pipe that is short and connected to a container that is clearly signed

▶ **Figure 2.15** Fire fighting signs
  a fire extinguisher
  b fire hose
  c supplementary green/white 'This way' signs for fire fighting equipment

▶ **Figure 2.16** Emergency escape signs (note that a different style of 'running man' is given in BS 5499, but the running man complies with Regulations)

▶ a container where the contents may change regularly; for example, chemical process pipework. In such cases employers need to make other arrangements for ensuring employees know the dangerous properties of the container. Employers could provide suitable process instruction sheets or training for employees.

The labels on containers or pipes containing dangerous substances should be in accordance with *Directive 67/584/EEC* and *88/379/EEC* or as per the warning signs in the general descriptions. Such warning signs can be supplemented by additional information. When the containers are being transported they need to be supplemented by suitable signs. All signs must be mounted on the visible side in non-pliable self-adhesive or painted form. Labels used on pipes must be positioned visibly in the vicinity of the most dangerous points, such as valves and joints, and at reasonable intervals. Areas, rooms or enclosures used for the storage of dangerous substances or preparations must be marked and labelled.

▶ **Figure 2.17** Warning signs for pipes

2.4.4 Hand signals
The regulations recognise and have requirements for acoustic signals, verbal communications and hand signals. Generally accepted hand signals are indicated in Figure 2.18. Refer to the HSE publication *Safety signs and signals* (L64) for additional signals for horizontal movements.

2.4.5 Identification and notices for electrical installations
BS 7671 gives a list of requirements for identification and notices for electrical installations that are important to those responsible for the operation and maintenance of the installation.

For example, except where there is no possibility of confusion, a label or other means of identification is required to be provided to indicate the purpose of each item of

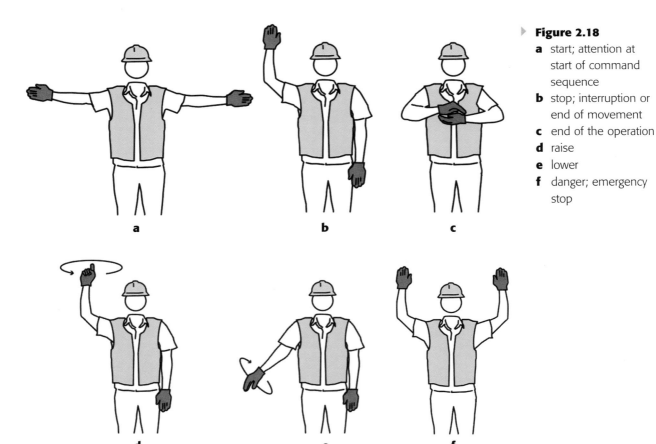

▶ **Figure 2.18**
 a start; attention at start of command sequence
 b stop; interruption or end of movement
 c end of the operation
 d raise
 e lower
 f danger; emergency stop

switchgear and controlgear. This requirement is reinforced by the *Electricity at Work Regulations 1989*, and is essential for safe operation. Regulation 514-01-02 of BS 7671 goes further, and states that, as far as is reasonably practicable, wiring shall be arranged or marked so that it can be identified for inspection, testing, repair or alteration of the installation. This means that, for example, cable runs from switchboards should be grouped together. Cables and conductors should rung neatly and logically on cable trays etc.

▶ **Table 2.8**
Requirements for
identification in BS 7671

The requirements for identification in BS 7671 are summarised in Table 2.8.

Requirements for identification in BS 7671		Regulation
Area	areas reserved for skilled or instructed persons	473-13-03
Buried cables	cables buried in the ground, cable covers, marking tape, identification of conduits and ducts	522-06-03
Caravans	electrical inlet indicating nominal voltage, frequency and rated current	608-07-03
	instructions for electricity supply and periodic inspection and testing	608-07-05
Earthing and bonding	warning – earthing and bonding connections	514-13-01 542-03-03
Highway power supply	identification of highway power supply cables	611-04-03
	temporary highway supply unit (identification of maximum sustained current)	611-06-02
Inspection and testing	periodic inspection and testing	514-12-01
	quarterly test of RCDs	514-12-02
Isolation and switching	isolating device, identification of the installation or circuit	461-01-05 537-02-09
	identification of device for switching for mechanical maintenance	462-01-02 537-03-02
	identification of emergency switching device	463-01-04
	identification of fireman's switch	476-03-07 537-04-06
Live parts	live parts not capable of being isolated by a single device	461-01-03 514-11-01
Protective conductor current	warning – circuits with high protective conductor current	607-03-02
Protective measures	warning – earth-free equipotential bonding	471-11-01 514-13-02
	warning – electrical separation	471-12-01 514-13-02

Requirements for identification in BS 7671 *continued*		Regulation
Switchgear and control gear	purpose of switchgear and controlgear	514-01-01
	identification of protective devices	514-08-01
	overcurrent protective device, type of fuse link, nominal current of fuse or circuit-breaker	533-01-01 533-01-02
	switchboard busbar or conductor	514-03-03
Voltage	voltages in excess of 230 V prior to gaining access	514-10-01
	mixed voltages, identification on means of access of voltages present	514-10-01
Wiring and conductors	identification of wiring	514-01-02
	identification of conduit	514-02-01
	identification of cable cores	514-03-01 514-03-02
	identification of conductors by colour	514-04
	identification of conductors by letters and/or numbers	514-05
	omission of identification by colour or marking	514-06
	wiring to both 'old' and 'new'colours	514-14-01
	diagrams, charts	514-09-01

2.4.6 Identification of conductors by colour

Table 51 from BS 7671 gives requirements for the identification of conductors by colour (Table 2.9).

▶ **Table 2.9** Identification of conductors (from Table 51 in BS 7671)

Function	Alpha-numeric	Colour
Protective conductors		green-and-yellow
Functional earthing conductor		cream
a.c. power circuit[1]		
Phase of single-phase circuit	L	brown
Neutral of single- or three-phase circuit	N	blue
Phase 1 of three-phase a.c. circuit	L1	brown
Phase 2 of three-phase a.c. circuit	L2	black
Phase 3 of three-phase a.c. circuit	L3	grey
Two-wire unearthed d.c. power circuit		
Positive of two-wire circuit	L+	brown
Negative of two-wire circuit	L−	grey
Two-wire earthed d.c. power circuit		
Positive (of negative earthed) circuit	L+	brown
Negative (of negative earthed) circuit[2]	M	blue
Positive (of positive earthed) circuit[2]	M	blue
Negative (of positive earthed) circuit	L−	grey
Three-wire d.c. power circuit		
Outer positive of two-wire circuit derived from three-wire system	L+	brown
Outer negative of two-wire circuit derived from three-wire system	L−	grey
Positive of three-wire circuit[3]	L+	brown
Mid-wire of three-wire circuit[2,3]	M	blue
Negative of three-wire circuit	L−	grey
Control circuits, ELV and other applications		
Phase conductor	L	brown, black, red, orange, yellow, violet, grey, white, pink or turquoise
Neutral or mid-wire[4]	N or M	blue

Notes:

1 power circuits include lighting circuits
2 M identifies either the mid-wire of a three-wire d.c. circuit, or the earthed conductor of a two-wire earthed d.c. circuit
3 only the middle wire of three-wire circuits may be earthed
4 an earthed PELV conductor is blue

2.4.7 Conduit and other pipes

Where conduit is to be distinguished from pipes or other services, orange is the colour to be used. The colour identifiers for other services are:

Recommended safety colour	BS colour reference BS 4800	Purpose	Example
Red	04 E 53	fire fighting	
Yellow	08 E 51	warning	
Auxiliary blue	18 E 53	with basic identification colour green – fresh water (potable or non-potable)	

▸ **Table 2.10** Safety colours

Pipe contents	Basic identification colour names[2]	BS identification colour reference BS 4800[1]	Example
Water	Green	12 D 45	
Steam	Silver-grey	10 A 03	
Oils – mineral, vegetable or animal Combustible liquids	Brown	06 C 39	
Gases in either gaseous or liquefied condition (except air)	Yellow ochre	08 C 35	
Acids and alkalis	Violet	22 C 37	
Air	Light blue	20 E 51	
Other liquids	Black	00 E 53	
Electrical services and ventilation ducts	Orange	06 E 51	

▸ **Table 2.11** Basic identification of pipe colours

Notes:

1 some colours are marginally outside the limits specified in ISO/R 508 but for practical purposes they may be used

2 the colour names given in column 2 are only included for guidance since different colour names may be used by different manufacturers for the same colour reference

Electrical installations

3.1 The need for maintenance

Regulation 4(2) of the *Electricity at Work Regulations 1989* requires that: 'As may be necessary to prevent danger, all systems shall be maintained so as to prevent, so far as is reasonably practicable, such danger.'

The *Memorandum of Guidance* published by the Health and Safety Executive advises that this regulation is concerned with the need for maintenance to ensure the safety of the system, rather than being concerned with the activity of doing the maintenance in a safe manner (which is required by regulation 4(3)). The obligation to maintain arises if danger would otherwise result. There is no specific requirement to maintain as such; what is required is that the system be kept in a safe condition. The quality and frequency of the maintenance must be sufficient to prevent danger, so far as is reasonably practicable. The HSE *Memorandum* advises that regular inspection of equipment is an essential part of any preventive maintenance programme. Practical experience of use of the installation may indicate an adjustment to the frequency at which preventive maintenance is to be carried out. This is a matter for the judgement of the duty holder who should seek all the information he needs to make this judgement including advice from equipment manufacturers.

▶ **Figure 3.1** HSR 25: *Memorandum of Guidance on the EWR*

3.2 Fixed installations, equipment and appliances

Regulation 4(2) of the *Electricity at Work Regulations* makes reference to all systems being maintained so as to prevent, so far as is reasonably practicable, such danger.

Systems are defined as follows: '"System" means an electrical system in which all the electrical equipment is, or may be electrically connected to a common source of electrical energy and includes such source and such equipment' (regulation 2(1)).

Equipment, mentioned in the above definition, is defined as: '"Electrical equipment" includes anything used, intended to be used or installed for use, to generate, provide, transmit, transform, rectify, convert, conduct, distribute, control, store, measure or use electrical energy' (regulation 2(1)).

As a consequence of these definitions, 'system' includes all electrical equipment from the generating equipment and the distribution network to the fixed wiring of a building, and all the equipment in the building including fixed equipment and portable and hand-held appliances. Electrical equipment includes anything powered by whatever source of electrical energy including battery-powered.

Figure 3.2 BS 7671: 2001 (2004) *Requirements for Electrical Installations*

The scope of this section of *Electrical Maintenance* can be considered to include:

▸ distribution systems
▸ electrical installations (note that electrical installations are not always within buildings)
▸ electrical equipment supplied from electrical installations.

There is generally a distinction drawn between the fixed electrical installation of the building covered by BS 7671: 2001 (2004) *Requirements for Electrical Installations* and other items supplied from the fixed installation including appliances. The inspection, testing and maintenance requirements of the fixed electrical installation as covered by BS 7671 are discussed later in this publication. The in-service inspection and testing of other electrical equipment including appliances is discussed in Chapter 6.

3.3 Frequency of inspection and test of fixed installations

The *Health and Safety at Work etc. Act 1974* states, in Section 6, that:

> It shall be the duty of any person who designs, manufactures, imports or supplies any article for use at work to take such steps as are necessary to secure that there will be available in connection with the use of the article at work adequate information about the use for which it is designed and has been tested, and about any conditions necessary to ensure that, when put to use, it will be safe and without risks to health.

There is a clear duty upon designers and installers of electrical installations to provide information on the need or otherwise for inspection and testing and its frequency. When an installation is designed and installed, assumptions are made by the designer and installer as to what is likely to be the use and abuse of a system. The designer will have assumed certain intervals between inspections, and inspections and tests, in his design. The Electrical Installation Certificate of Appendix 6 of BS 7671 requires that the interval at which the installation must be inspected and tested be inserted.

The interval inserted on the certificate is an initial maximum period. Experience will show whether the interval can be extended or needs to be shortened. In the absence of guidance from the original designer and installer, minimum periods between initial inspections are given in Table 3.1.

3.4 Inspection and testing regimes

The *Management of Health and Safety at Work Regulations* require employers to give consideration as to how they are going to manage health and safety matters. This applies to electrical installations as much as to any other safety matter. Detailed inspection and testing, however thorough and expensive, say every five years, is not going to ensure the continuing safety of an electrical installation which might suffer damage on a daily basis. Consideration needs to be given to a maintenance regime. General domestic installations have a recommended maximum period between inspections of ten years, but it is presumed that the householder will naturally identify any faults and breakages and arrange to have them repaired in the periods between inspections. The periods between inspection given in the third column of Table 3.1 are obviously too long if defects are not rectified in between times. These inspections are

Type of installation	Routine check	Maximum period between inspections and testing as necessary	Reference (see notes below)
General installations			
Domestic	—	change of tenancy/10 years	—
Commercial	1 year	change of tenancy/5 years	1, 2
Educational establishments	4 months	5 years	1, 2
Hospitals	1 year	5 years	1, 2
Industrial	1 year	3 years	1, 2
Residential accommodation	at change of occupancy/1 yr	5 years	1
Offices	1 year	5 years	1, 2
Shops	1 year	5 years	1, 2
Laboratories	1 year	5 years	1, 2
Buildings open to the public			
Cinemas	1 year	3 years	2, 6, 7
Church installations	1 year	5 years (quinquennially)	2
Leisure complexes[a]	1 year	3 years	1, 2, 6
Places of public entertainment	1 year	3 years	1, 2, 6
Restaurants and hotels	1 year	5 years	1, 2, 6
Theatres	1 year	3 years	2, 6, 7
Public houses	1 year	5 years	1, 2, 6
Village halls/Community centres	1 year	5 years	1, 2
Special installations			
Agricultural and horticultural	1 year	3 years	1, 2
Caravans	1 year	3 years	—
Caravan parks	6 months	1 year	1, 2, 6
Highway power supplies	as convenient	6 years	—
Marinas	4 months	1 year	1, 2
Fish farms	4 months	1 year	1, 2
Swimming pools	4 months	1 year	1, 2, 6
Emergency lighting	daily/monthly	3 years	2, 3, 4
Fire alarms	daily/weekly/monthly	1 year	2, 4, 5
Launderettes	1 year	1 year	1, 2, 6
Petrol filling stations	1 year	1 year	1, 2, 6
Construction site installations	3 months	3 months	1, 2

Reference key

1 Particular care must be taken to comply with S1 2002 No. 2665 the *Electricity Safety, Quality and Continuity Regulations 2002*.

2 S1 1989 No. 635 the *Electricity at Work Regulations 1989 (Regulation 4 and Memorandum)*.

3 See BS 5266: 1999 *Emergency lighting Part 1 Code of practice for the emergency lighting of premises other than cinemas and certain other specified premises used for entertainment*.

4 Other intervals are recommended for testing the operation of batteries and generators.

5 See BS 5839: 2002 *Fire detection and fire alarm systems for buildings Part 1 Code of practice for system design installation, commissioning and maintenance*.

6 Local authority conditions of licence.

7 S1 1995 No. 1129 (Clause 27) the *Cinematography (Safety) Regulations*.

a excluding swimming pools

▶ **Table 3.1** Frequencies of periodic inspection and testing of electrical installations

Figure 3.3 Many defects will be easily apparent

to determine if there has been deterioration in the installation, and whether changes are necessary to bring the installation into line with the current standard.

In the workplace it may not be reasonable to expect routine reporting of defects, so regular routine checks must be carried out. Routine checks are performed to identify breakage and wear. Breakages and deterioration due to wear and tear cannot be left for such periods. In areas open to the public where defects might not be reported, further checks (see the second column of Table 3.1) must supplement the inspections of the third column.

The frequency of routine checks will depend upon the circumstances, and they do not necessarily need to be carried out by electrically skilled staff. Frequent, even daily checks may be appropriate, particularly in areas open to the public. Table 3.2 summarises the activities of a routine check and the defects looked for.

Table 3.2 Routine checks

Activity	Check	
Check defect reports	all reported defects have been rectified	
Inspection	look for: ▶ breakages ▶ wear or deterioration ▶ signs of overheating ▶ missing parts (covers, screws) ▶ switchgear accessible (not obstructed) ▶ loose fixings ▶ doors of enclosures secure ▶ adequate labelling.	
Operation	operate: ▶ switchgear (where reasonable) ▶ equipment – switch off and on ▶ Residual Current Devices (RCDs) by using the test button	

3.5 Periodic inspection and testing

The requirements for periodic inspection and testing are provided in Chapter 73 of BS 7671. It is worth considering exactly what the requirements are. Regulations 731-01-02, 731-01-03, 731-01-04 and 731-01-05 state:

Regulation 731-01-02: Periodic inspection and testing of an electrical installation shall be carried out to determine, so far as is reasonably practicable, whether the installation is in a satisfactory condition for continued service.

Regulation 731-01-03: Inspection comprising careful scrutiny of the installation shall be carried out without dismantling or with partial dismantling as required, together with the appropriate tests of Chapter 71. The scope of the periodic inspection and testing shall be decided by a competent person, taking into account the availability of records and the use, condition and nature of the installation.

Regulation 731-01-04: Such inspection and testing shall provide, so far as is reasonably practicable, for:

i the safety of persons and livestock against the effects of electric shock and burns, in accordance with regulation 130-01, and

ii protection against damage to property by fire and heat arising from an installation defect, and

iii confirmation that the installation is not damaged or deteriorated so as to impair safety, and

iv the identification of installation defects and non-compliance with the requirements of the Regulations which may give rise to danger.

Regulation 731-01-05: Precautions shall be taken to ensure that the inspection and testing does not cause danger to persons or livestock and does not cause damage to property and equipment, even if the circuit is defective.

The requirements are for careful scrutiny (inspection) supplemented by appropriate testing as necessary. The intent is that, where possible, the installation should not be dismantled as this obviously introduces the risk of it not being correctly reassembled, and that any partial dismantling be carried out only as required. The careful scrutiny is to be 'supplemented' by appropriate testing from Chapter 71 to verify compliance with the general requirements for safety. The general requirements for safety are summarised in the four parts of regulation 731-01-04.

Where possible, the general condition of an installation should be assessed by the electrical contractor before the detailed work of inspection and testing is started. Agreement can then be reached prior to the work proper being commenced as to the testing likely to be necessary. This is obviously preferable to differences arising after completion of the work as to what was contracted to be carried out.

Often, inspection and test work is tendered for competitively, when there need to be good guidelines on what proportion of the installation is to be tested. To assist in these situations, a typical requirement is listed in Table 3.3. The guidance in Table 3.3 applies to installations where no alterations are known to have been made since the last inspection and test.

When a periodic inspection and test is carried out, the periodic inspection report of Appendix 6 of BS 7671 should be completed by the person carrying out the inspection and testing, together with an installation schedule including test results as found in *IEE Guidance Note 3*.

▶ **Figure 3.4** *IEE Guidance Note 3* gives detailed information on the inspecting and testing of electrical installations

3.6 Certificates

The certificates of BS 7671 are included in an appendix to the Standard. The certificates used do not have to be identical to those in BS 7671. It may be appropriate to use a different layout. For example, an electrician carrying out many minor repairs in a day may well use a works order form that allows a number of minor electrical works to be recorded on the one certificate.

▶ **Table 3.3** Summary of periodic testing to be performed on existing installations [1, 2, 3]

Test type	Recommendation
Continuity of circuit protective conductors	Tests to be carried out between the earth terminal or main earthing terminal (MET) of the distribution board or consumer unit and the following exposed-conductive-parts: ▶ socket-outlet earth connections ▶ accessible exposed-conductive-parts of current-using equipment and accessories. The continuity of the earthing conductor must be checked
Continuity of equipotential bonding conductors	Continuity must be established for: ▶ main equipotential bonding conductors ▶ supplementary equipotential bonding conductors.
Continuity of ring final circuit conductors	All conductors, including circuit protective conductors, forming part of a ring final circuit must be tested
Insulation resistance	Insulation resistance between live conductors and between live conductors and earth should be measured. Equipment which may be vulnerable to an insulation test should first be disconnected. For a 230 V installation, insulation tests should be performed at 500 V d.c. with a test apparatus capable of delivering 1 mA and the minimum value of insulation resistance should be 0.5 MΩ.
Polarity	Tests to be carried out at: ▶ origin of the installation ▶ distribution boards ▶ fuses and single-pole control and protective devices ▶ socket-outlets and similar accessories ▶ centre-contact lamp holders.
Earth electrode resistance	Test each earth rod or group of rods separately, with the test links removed, and with the installation isolated from the supply source.
Earth fault loop impedance [4]	Tests to be carried out at: ▶ the origin of the installation ▶ each distribution board ▶ each socket-outlet ▶ the extremity of every radial circuit.
Functional [5]	Test to be carried out: ▶ circuit-breakers, isolation and switching devices to be manually operated to verify the devices disconnect the supply ▶ operation of all RCDs to be verified by a suitable test instrument and by the operation of the test button ▶ labels to be checked.

Notes:

1 The person carrying out the testing is required to decide which of the above tests are appropriate by using his/her experience and knowledge of the installation being inspected and tested and by consulting any available records.

2 Where sampling is applied, the percentage used is at the discretion of the tester. However, a percentage of less than 10% is inadvisable.

3 The tests need not be carried out in the order shown in the table.

4 The earth fault loop impedance test may be used to confirm the continuity of protective conductors at socket-outlets and at accessible exposed-conductive-parts of current-using equipment and accessories.

5 Some earth loop impedance testers may trip residual current devices in the circuit.

Form F1

Form No 123/1

ELECTRICAL INSTALLATION CERTIFICATE (notes 1 and 2)

(REQUIREMENTS FOR ELECTRICAL INSTALLATIONS - BS 7671 [IEE WIRING REGULATIONS])

▶ **Figure 3.5** Electrical Installation Certificate

DETAILS OF THE CLIENT (note 1)	House Builder Ltd, 1 City Way, LONDON		

INSTALLATION ADDRESS Plot 24, New Road NEW TOWN County   Postcode ABT 2CD

DESCRIPTION AND EXTENT OF THE INSTALLATION Tick boxes as appropriate

Description of installation: Domestic

Extent of installation covered by this Certificate:

Complete electrical, including smoke and intruder alarms

(Use continuation sheet if necessary)   see continuation sheet No: ....

New installation ☑

Addition to an existing installation ☐

Alteration to an existing installation ☐

FOR DESIGN, CONSTRUCTION, INSPECTION & TESTING
I being the person responsible for the Design, Construction, Inspection & Testing of the electrical installation (as indicated by my signature below), particulars of which are described above, having exercised reasonable skill and care when carrying out the Design, Construction, Inspection & Testing, hereby CERTIFY that the said work for which I have been responsible is to the best of my knowledge and belief in accordance with BS 7671: .2001.., amended to ..2004.. (date) except for the departures, if any, detailed as follows:

Details of departures from BS 7671 (Regulations 120-01-03, 120-02):

None

The extent of liability of the signatory is limited to the work described above as the subject of this Certificate.

Name (IN BLOCK LETTERS): A SMITH
Signature (note 3): *a Smith*
For and on behalf of: All Electrics Ltd
Address: 27, Central Road
NEW TOWN
County   Postcode...EF3 4GH

Position: Director
Date: 20/4/2005

Tel No: ....

NEXT INSPECTION
I recommend that this installation is further inspected and tested after an interval of not more than ..10...... years/months. (notes 4 and 7)

SUPPLY CHARACTERISTICS AND EARTHING ARRANGEMENTS Tick boxes and enter details, as appropriate

Earthing arrangements	Number and Type of Live Conductors	Nature of Supply Parameters	Supply Protective Device Characteristics
TN-C ☐ TN-S ☐ TN-C-S ☑ TT ☐ IT ☐	a.c. ☑ d.c. ☐ 1-phase, 2-wire ☑ 2-pole ☐ 1-phase, 3-wire ☐ 3-pole ☐ 2-phase, 3-wire ☐ other ☐ 3-phase, 3-wire ☐ 3-phase, 4-wire ☐	Nominal voltage, U/U_0 (1) ...230... V Nominal frequency, f (1) .....50...Hz Prospective fault current, I_{pf} (2) .16... kA (note 6) External loop impedance, Z_e (2) 0.35Ω (Note: (1) by enquiry, (2) by enquiry or by measurement)	Type: BS 1361 fuse Nominal current rating ...100....A
Alternative source ☐ of supply (to be detailed on attached schedules)			

Page 1 of 4 (note 5)

PARTICULARS OF INSTALLATION REFERRED TO IN THE CERTIFICATE Tick boxes and enter details, as appropriate

Means of Earthing	Maximum Demand
Distributor's facility ☑	Maximum demand (load) ....60...... Amps per phase
Installation earth electrode ☐	**Details of Installation Earth Electrode** (*where applicable*)

	Type (e.g. rod(s), tape etc)	Location	Electrode resistance to earth
	.......None..........	....	.... Ω

Main Protective Conductors

Earthing conductor: material ......Copper.... csa ....16........mm²   connection verified ☑

Main equipotential bonding conductors   material ......Copper.... csa ....10.......mm²   connection verified ☑

To incoming water and/or gas service ☑   To other elements: ....

Main Switch or Circuit-breaker

BS, Type ...BS EN 61009.......   No. of poles ....2......   Current rating ....80..A   Voltage rating ....230..V

Location ....Garage............................   Fuse rating or setting.............——......A

Rated residual operating current $I_{\Delta n}$ = ...30. mA, and operating time of 200 ms (at $I_{\Delta n}$) (applicable only where an RCD is suitable and is used as a main circuit-breaker)

COMMENTS ON EXISTING INSTALLATION (in the case of an alteration or additions see Section 743):

..........New installation...

....

....

SCHEDULES (note 2)

The attached Schedules are part of this document and this Certificate is valid only when they are attached to it.
.....1...... Schedule(s) of Inspections and ...1........ Schedule(s) of Test Results are attached.
(Enter quantities of schedules attached).

Page 2 of 4 (note 5)

Form F3   Form No 123 /3

SCHEDULE OF INSPECTIONS

Methods of protection against electric shock

(a) Protection against both direct and indirect contact:

- [N/A] (i) SELV (note 1)
- [N/A] (ii) Limitation of discharge of energy (note 2)

(b) Protection against direct contact: (note 3)

- [✓] (i) Insulation of live parts
- [✓] (ii) Barriers or enclosures
- [N/A] (iii) Obstacles (note 4)
- [N/A] (iv) Placing out of reach (note 5)
- [N/A] (v) PELV (note 6)
- [✓] (vi) Presence of RCD for supplementary protection

(c) Protection against indirect contact:

- (i) EEBAD (note 7) including:
- [✓] Presence of earthing conductor
- [✓] Presence of circuit protective conductors
- [✓] Presence of main equipotential bonding conductors
- [✓] Presence of supplementary equipotential bonding conductors
- [N/A] Presence of earthing arrangements for combined protective and functional purposes (note 8)
- [N/A] Presence of adequate arrangements for alternative source(s), where applicable
- [✓] Presence of residual current device(s)
- [N/A] (ii) Use of Class II equipment or equivalent insulation (note 9)
- [N/A] (iii) Non-conducting location: (note 10) Absence of protective conductors
- [N/A] (iv) Earth-free equipotential bonding: (note 11) Presence of earth-free equipotential bonding conductors
- [N/A] (v) Electrical separation (note 12)

Prevention of mutual detrimental influence

- [✓] (a) Proximity of non-electrical services and other influences
- [✓] (b) Segregation of band I and band II circuits or band II insulation used
- [✓] (c) Segregation of safety circuits

Identification

- [✓] (a) Presence of diagrams, instructions, circuit charts and similar information
- [✓] (b) Presence of danger notices and other warning notices
- [✓] (c) Labelling of protective devices, switches and terminals
- [✓] (d) Identification of conductors

Cables and conductors

- [✓] (a) Routing of cables in prescribed zones or within mechanical protection
- [✓] (b) Connection of conductors
- [✓] (c) Erection methods
- [✓] (d) Selection of conductors for current-carrying capacity and voltage drop
- [✓] (e) Presence of fire barriers, suitable seals and protection against thermal effects

General

- [✓] (a) Presence and correct location of appropriate devices for isolation and switching
- [✓] (b) Adequacy of access to switchgear and other equipment
- [✓] (c) Particular protective measures for special installations and locations
- [✓] (d) Connection of single-pole devices for protection or switching in phase conductors only
- [✓] (e) Correct connection of accessories and equipment
- [N/A] (f) Presence of undervoltage protective devices
- [✓] (g) Choice and setting of protective and monitoring devices for protection against indirect contact and/or overcurrent
- [✓] (h) Selection of equipment and protective measures appropriate to external influences
- [✓] (i) Selection of appropriate functional switching devices

Inspected by ...*A. Smith*......................   Date ....*20/4/2005*............................

Notes:

✓   to indicate an inspection has been carried out and the result is satisfactory

✗   to indicate an inspection has been carried out and the result was unsatisfactory

N/A   to indicate the inspection is not applicable

LIM   to indicate that, exceptionally, a limitation agreed with the person ordering the work prevented the inspection or test being carried out.

1. SELV An extra-low voltage system which is electrically separate from earth and from other systems. The particular requirements of the Regulations must be checked (see Regulations 411-02 and 471-02)

2. Limitation of discharge of energy - not adopted for domestic installations, used on appliances and equipment.

3. Method of protection against direct contact - will include measurement of distances where appropriate.

4. Obstacles - not suitable for domestic installations, only adopted in special circumstances (see Regulations 412-04 and 471-06)

5. Placing out of reach - not suitable for domestic installations, only adopted in special circumstances (see Regulations 412-05 and 471-07)

6. PELV An extra-low voltage system which is electrically separate from other systems but not earth. The particular requirements of the Regulations must be checked (see Regulations 411-02 and 471-14)

7. EEBAD Earthed equipotential bonding and automatic disconnection of supply, the common form of indirect shock protection

8. Combined protective and functional earthing - it is normal to combine protective and functional earthing. In non-domestic systems functional earthing of IT systems may be separated, (clean earth).

9. Use of Class II equipment - not suitable for domestic installations, infrequently adopted and only when the installation is to be supervised (see Regulations 413-03 and 471-09)

10. Non-conducting locations - not suitable for domestic installations and requiring special precautions (see Regulations 413-04 and 471-10)

11. Earth-free local equipotential bonding - not suitable for domestic installations, only used in special circumstances (see Regulations 413-05 and 471-14)

12. Electrical separation - not adopted in domestic installations (see Regulations 413-06 and 471-12)

Page 3 of 4

© The Institution of Engineering and Technology

Form 4
SCHEDULE OF TEST RESULTS

Form No 123 /4

Contractor: ...... All Electrics Ltd ....
Test Date: .... 20/4/2005 ....
Signature .... *A. Smith*
Method of protection against indirect contact: .E. E. B. A. D. S.
Equipment vulnerable to testing: .30 mA RCDs circuits 7 and 4, dimmer and fluorescent circuit 2, Shower circuit 6 ....

Address/Location of distribution board:
.... Plot 24, New Road ....
.... Town ....
.... County ....

* Type of Supply: TN-S/TN-C-S/TT~
* Ze at origin: 0.35 ohms
* PFC: ...... 16 .. kA

Instruments
loop impedance: ...... AB 11 ....
continuity: .... AB 22 ....
insulation: ...... AB 44 ....
RCD tester: ... AB 55 ....

Description of Work: ..House electrical installation..

Circuit Description	Overcurrent Device * Short-circuit capacity: 6.kA		Wiring Conductors		Test Results										
					Continuity			Insulation Resistance		P o l a r i t y	Earth Loop Imped-ance Z_s	Functional Testing		Remarks	
	type	Rating I_n	live	cpc	$R_1 + R_2$	R_2	R i n g	Live/ Live	Live/ Earth			RCD time	Other		
		A	mm²	mm²	Ω	Ω		MΩ	MΩ		Ω	ms			
1	2	3	4	5	*6	*7	*8	*9	*10	*11	*12	*13	*14		15
1 Lights up	B	10	1.5	1.0	2.4	—	—	50	40	✓	2.8	—	✓		
2 Lights down	B	10	1.5	1.0	2.7	—	—	—	30	✓	3.1	—	✓	Dimmer
3 Sockets up	B	32	2.5	1.5	0.4	0.3	✓	30	30	✓	0.8	—	✓	
4 Sockets down	B	32	2.5	1.5	0.5	0.3	✓	—	30	✓	0.9	200	✓	RCD, Vulnerable
5 Cooker	B	32	6.0	2.5	0.1	—	—	50	40	✓	0.5	—	✓	
6 Shower	B	45	10.0	4.0	0.15	—	—	—	40	✓	0.5	200	✓	Electronic
7 Garage	B	20	2.5	1.5	0.4	—	—	—	30	✓	0.8	200	✓	RCD

Deviations from Wiring Regulations and special notes:

None

* See notes on schedule of test results

Testing

4

The following tests, where relevant, including measurements where specified, are to be carried out on fixed low voltage electrical installations as required by BS 7671 during initial and periodic inspections and after alteration, addition, repair or modification:

▶ continuity of conductors including protective conductors and bonding conductors
▶ insulation resistance
▶ polarity
▶ earth electrode resistance
▶ earth fault loop impedance
▶ prospective fault current
▶ functional testing including the operation of RCDs.

Electrical installations must be securely isolated and proven dead before testing is performed with the exception of certain polarity tests, loop impedance, prospective fault current and RCD tests.

Further detailed guidance on testing electrical installations is given in *IEE Guidance Note 3: Inspection and testing*.

4.1 Continuity

Continuities may be tested by either Test Method 1 (install a temporary shorting link at the distribution board – Figure 4.1) or Test Method 2 (use a flying lead – Figure 4.2).

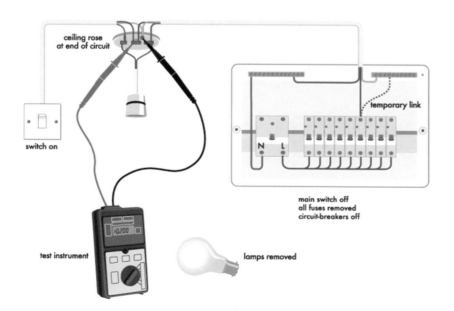

▶ **Figure 4.1** Test Method 1 – temporary shorting link installed at distribution board

4

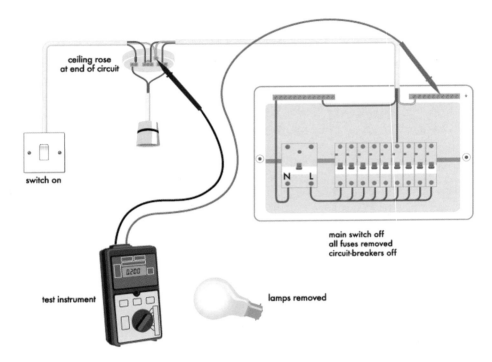

Figure 4.2 Test Method 2 – use of a flying lead

ceiling rose at end of circuit

switch on

main switch off
all fuses removed
circuit-breakers off

test instrument

lamps removed

Continuity testing of conductors should be carried out with an instrument having a no-load voltage of between 4 and 24 V d.c. or a.c. and a short-circuit current of not less than 200 mA. This aligns the requirement of BS 7671 with the CENELEC *Harmonisation Document* (HD 384.6.61 S1). The resistance of the leads of the test instrument should be measured and subtracted from the readings.

Where a ferrous enclosure such as a steel conduit or trunking is used as the protective conductor, particular care must be exercised when testing, as a poor contact between sections may nonetheless give a satisfactory test result. It is recommended that the following procedure be followed:

1 Inspect the enclosure along its length, checking all the joints for electrical and mechanical soundness.
2 Carry out a standard continuity (ohmmeter) test, using Test Method 2, Figure 4.2.
3 If there are concerns regarding the soundness of the conduit or trunking to act as a protective conductor, a high current test may be necessary. A.c. ohmmeters with 50 V output are available that can be used to test at up to 1.5 times the design current up to a maximum of 25 A. Care needs to be exercised since sparking can occur at loose joints and there can be a risk of electric shock both for the person performing the test and other persons who may be present.

4.2 Insulation resistance

Insulation measurement for 230/400 V systems must be made at 500 V d.c. and the test instrument must be capable of supplying the test voltage when loaded with 1 mA. That is, the insulation tester must be able to maintain 500 V when loaded with the minimum acceptable insulation resistance of 0.5 MΩ.

Although an insulation resistance of not less than 0.5 MΩ complies with BS 7671, where a reading of less than 2 MΩ is recorded on a distribution board (with all circuits connected) the possibility of a defect exists. Then, each circuit should be individually tested and the measured insulation resistance of each should exceed 2 MΩ.

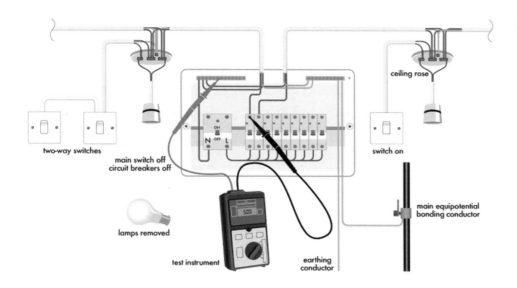

Electrical equipment cannot function properly if its electrical insulation is not in good condition. Generally, insulation deteriorates relatively slowly so that by regular monitoring action can be taken before the deterioration becomes critical.

The insulation resistance of materials is affected by many factors – moisture, contamination, oil, corrosive substances, vibration, heat and ageing. For insulating material that can absorb water, the water content is probably the most important factor in determining insulation readings. Many of the effects on insulation are reversible so that insulation readings may depend upon the type of weather, whether the machine or equipment is at running temperature etc. The insulation of mineral insulated cooker elements will be relatively low if the elements have absorbed moisture. After, say, half an hour of use, the moisture will be driven off and the insulation reading will be considerably higher.

4.3 Polarity

Polarity checking is carried out to confirm that all equipment is connected as necessary to the correct pole of the supply; for example, that switches and fuses will interrupt the phase and not the neutral of the supply. There is a requirement in BS 7671 to carry out continuity testing of an installation before the supply is connected, and this can be used to confirm that the polarity is correct and that all switches and fuses are in the phase conductor(s).

There are instruments that give an indication of polarity when the supply is live. The most common are test lamps and neon testers. It is to be noted that test lamps and neon indicators simply indicate that there is a potential difference.

Many phase–earth loop testers and RCD testers are fitted with neon lights to check for correct wiring in an installation prior to measurement of loop impedance and RCD tests. The indication is normally by neon, initiated by a sufficient voltage between the phase and neutral/earth of the installation. Again, these checks indicate a potential difference and they will not necessarily indicate that there is a reversed neutral earth connection or indicate incorrect polarity of the incoming supply.

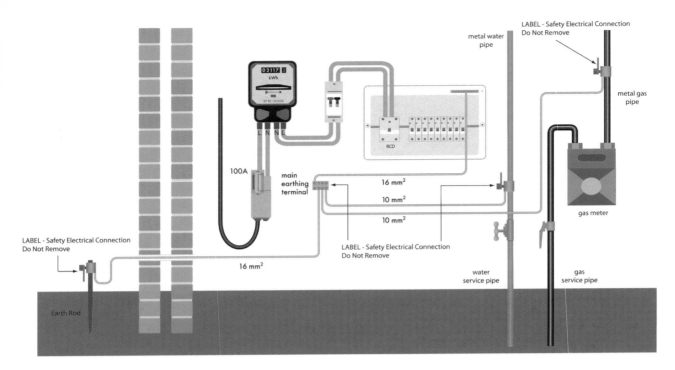

▶ **Figure 4.4** Installation forming part of a TT system with an earth electrode

4.4 Earth electrode

An installation forming part of a TT system will always be equipped with one or more earth electrodes as its means of earthing. The purpose of the earth electrode is to make good electrical contact with the general mass of earth in order to provide a suitable path for earth fault currents to return to the source. The source will, in most cases, be the distribution transformer which will have its own earth electrode. Inspection of the earth electrode and the earthing conductor and measurement of the resistance of the earth electrode or electrodes to the general mass of earth provides the information necessary to permit verifying that the installation design meets the requirements of BS 7671.

4.4.1 Types of earth electrode
An earth electrode generally consists of an earth rod or pipe, an earth tape or wire, a plate or underground structural metalwork embedded in foundations or other suitable underground metalwork (regulation 542-02-01 of BS 7671 refers).

The metallic structure of a steel-framed building which is embedded in concrete forming the foundations can form an excellent earth electrode with a resistance to earth as low as 1 Ω. However, the structural engineer should be consulted at an early stage as to whether the steelwork can be used as a foundation earth electrode because continuous earth currents may flow which could cause corrosion of the metalwork and cracking of the surrounding concrete. In addition, an adequate electrical connection will need to be made to the steelwork by means such as drilling and bolting.

4.4.2 Inspection of an earth electrode
The electrode should be located and inspected to ensure it is in good condition, properly connected, undamaged and fitted with a suitable label. The connecting conductor, the earthing conductor, must, once again, be properly routed where it will not suffer damage. Both the electrode and the cable, being external to the premises, are at risk from factors such as damage, corrosion and type and placement.

Damage

The design used, and the construction of an earth electrode, are such as to withstand damage and take account of the possible increase in resistance due to the effects of corrosion. Possible causes of damage may include (for example) vandalism, gardening, farm machinery, animals, or future excavation associated with building work.

Corrosion

Clause 11 of BS 7430: 1998 states that there are two aspects which should be considered regarding the corrosion resistance of an earth electrode or an earthing conductor. The first is compatibility with the soil itself, and the second is the possibility of galvanic (electrolytic) effects when an earth electrode is electrically connected to adjacent metalwork. Copper is one of the better and more commonly used materials for earth electrodes and underground conductors, but the corrosive effects of dissolved salts, organic acids and acid soils should generally be considered. Steel encased in concrete is generally protected against corrosion by the concrete. Where adjacent items of dissimilar buried metalwork are electrically connected together, the possibility of damage by galvanic (electrolytic) corrosion must be considered. Electrolytic corrosion could earth not only electrodes and earthing conductors but also cables, other underground services and structural metalwork.

An earth electrode must be able to pass the required current under fault conditions. The design of the electrode must be such that the heating effect of the energy dissipated (I^2t) to the soil will not result in a rise in the resistance of, or failure of, the earth electrode under both wet and dry conditions. An installation earth electrode for a TT system where automatic disconnection of the supply in the event of an earth fault is provided by a residual current device (RCD) is in almost all cases likely to be adequate.

Suitable type and placement

The location, type and embedded depth of the earth electrode should be such that soil freezing, in winter, or drying, in summer, will not result in an increase of earth electrode resistance above that necessary for safety.

4.4.3 Measurement of the resistance of an earth electrode

The resistance of an earth electrode may be measured by using either:

▸ a proprietary earth electrode test instrument or
▸ an earth fault loop impedance tester.

Earth electrode test instrument

The electrical installation must be securely isolated and the earth electrode should be disconnected at the electrode or at the main earthing terminal to avoid any parallel paths such as via the main equipotential bonding, and to ensure that all the test current passes through the earth electrode alone.

The installation will be unprotected against earth faults during the testing, which is why secure isolation is essential.

Testing should be carried out when the ground conditions are least favourable, such as during dry weather. The test requires the use of two temporary test spikes (electrodes), and is carried out in the following manner.

Connection to the earth electrode is made using terminals C1 and P1 of a four-terminal earth tester. To exclude the resistance of the test leads from the resistance reading,

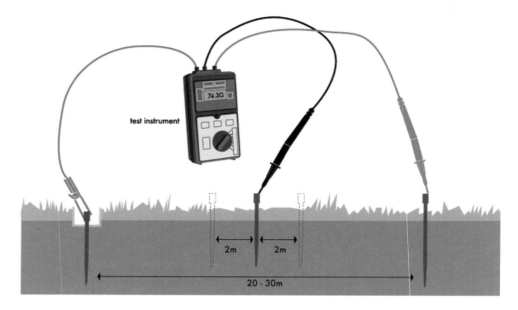

Figure 4.5 Earth electrode test – three-terminal instrument

individual leads should be taken from these terminals and connected separately to the electrode. If the test lead resistance is insignificant, the two terminals may be short-circuited at the tester and connection made with a single test lead, the same being true if using a three-terminal tester. Connection to the temporary spikes is made as shown in Figures 4.5 and 4.6.

The distance between the test spikes is important. If they are too close together, their resistance areas will overlap. In general, reliable results may be expected if the distance between the electrode under test and the current spike is at least ten times the maximum dimension of the electrode system, e.g. 30 m for a 3 m long rod electrode.

Three readings are taken: with the potential spike initially midway between the electrode and current spike; at a position 10 per cent of the electrode-to-current spike distance back towards the electrode; and at a position 10 per cent of the distance towards the current spike. By comparing the three readings, a percentage deviation can be determined. This is calculated by taking the average of the three readings, finding the maximum deviation of the readings from this average in ohms and expressing this as a percentage of the average.

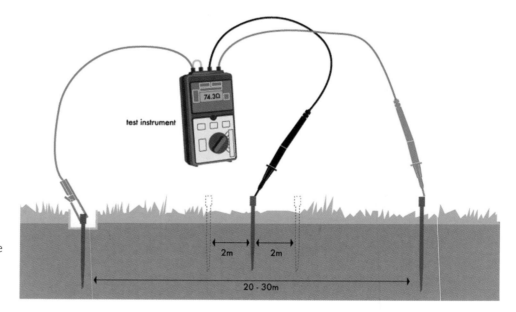

Figure 4.6 Earth electrode test – four-terminal instrument (configured to use three test leads)

The accuracy of the measurement using this technique is typically 1.2 times the percentage deviation of the readings. It is difficult to achieve a measurement accuracy better than 2 per cent, and inadvisable to accept readings that differ by more than 5 per cent. To improve the accuracy of the measurement to acceptable levels, the test must be repeated with a larger separation between the electrode and the current spike.

The instrument output current may be a.c. or reversed d.c. to overcome electrolytic effects. Because these testers employ phase sensitive detectors (PSDs), the errors associated with stray currents are eliminated.

The instrument should be capable of checking that the resistance of the temporary spikes used for testing is within the accuracy limits stated in the instrument specification. This may be achieved by an indicator provided on the instrument, or the instrument should have a sufficiently high upper range to enable a discrete test to be performed on the spikes.

If the temporary spike resistance is too high, measures to reduce the resistance will be necessary, such as driving the spikes deeper into the ground or watering with brine to improve contact resistance. In no circumstances should these techniques be used to temporarily reduce the resistance of the earth electrode under test.

After completion of the testing, the means of earthing for the installation must be reestablished before reenergising the installation.

Earth fault loop impedance tester

For measuring the resistance to earth of earth electrodes installed in domestic installations for use with an RCD, an earth fault loop impedance tester may be used. Before the test is undertaken the installation must be switched off and isolated and the earth conductor disconnected and, in addition, all equipotential bonding should be

▶ **Figure 4.7** Earth fault loop impedance tester

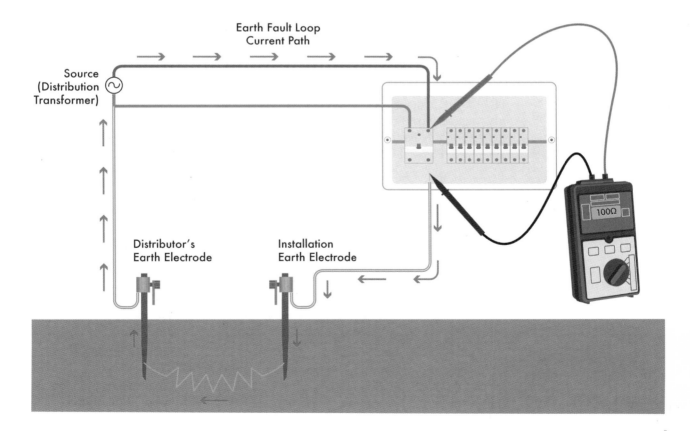

© The Institution of Engineering and Technology

disconnected from the earth electrode under test, to ensure that the test current passes through the earth electrode alone. The bonding must be reconnected after the test.

The loop tester is connected between the phase conductor at the origin of the installation and the earthing conductor connected to the earth electrode but disconnected from the main earthing terminal (for example by opening the test link). The impedance reading obtained is that of the complete phase–earth loop including the resistance of the earth electrode of the installation and the earth electrode of the distributor. The installation earth electrode resistance will be the major element of the resistance of this loop.

The earthing conductor and the main equipotential bonding must be reconnected after the test before the installation is reenergised.

Results of testing

The impedance reading taken is treated as the electrode resistance. BS 7671 requires:

$R_A I_{\Delta n} \leqslant 50$ V   for normal dry locations

$R_A I_{\Delta n} \leqslant 25$ V   for certain special locations such as construction sites and agricultural premises.

Where: R_A is the sum of the resistances of the earth electrode and the protective conductor(s) connecting it to the exposed-conductive-part, and $I_{\Delta n}$ is the rated residual operating current of the residual current device.

Maximum values of R_A for the basic standard ratings of residual current devices are given in Table 4.1, unless the manufacturer declares alternative values.

Table 4.1 indicates that the use of a suitably rated RCD will theoretically allow much higher values of R_A, and therefore of Z_s, than could be expected by using the overcurrent devices for indirect contact shock protection. In practice, however, values above 100 Ω will require further investigation.

BS 7430: *Code of practice for earthing* states that a value of earth electrode resistance exceeding 100 Ω may be unstable.

Where items of stationary equipment having a protective conductor current exceeding 3.5 mA in normal service are supplied from an installation forming part of a TT system,

Table 4.1 Maximum values of earth electrode resistance for TT installations

RCD rated residual operating current $I_{\Delta n}$ (mA)	Maximum value of earth electrode resistance, R_A (Ω)	
	normal dry locations	construction sites, agricultural and horticultural premises
30	1660	830
100	500	250
300	160	80
500	100	50

the product of the total protective conductor current (in amperes) and twice the resistance of the installation earth electrode(s) (in ohms) must not exceed 50.

The measured value and the method of measurement need to be recorded together with the approximate location of the electrode.

4.4.4 Methods for improving the resistance of an earth electrode

In some circumstances, it may be necessary to improve (reduce) the resistance of an earth electrode. For example, the resistance of a newly installed earth electrode may be found to be unacceptably high for coordination with the relevant protective devices, such as residual current devices. Some of the more commonly used methods for improving resistance follow.

Burying the earth electrode deeper into the soil

It may be possible to achieve some reduction in the resistance of an earth electrode by burying it deeper into the soil (e.g., in the case of an extensible rod-type electrode, by adding sections and driving it deeper).

The use of additional earth electrodes

Where the resistance of a single electrode is unacceptably high, a number of such electrodes may be connected in parallel. Where a number of vertical rod or pipe earth electrodes are so connected, the combined resistance is then practically proportional to the reciprocal of the number employed, provided that each is situated outside the resistance area of any other. It is often assumed that such rods or pipes are outside of each other's resistance areas if the mutual separation distance is not less than the driven depth. The clause also mentions that little is to be gained by a separation beyond twice the driven depth

Soil treatment or replacement

In special or difficult locations, the contact resistance of earth electrodes may be improved by means of soil replacement or treatment. Soil replacement involves replacing the soil in the immediate vicinity of the earth electrode with a low resistivity material, such as bentonite, conductive concrete or conductive cement. Soil treatment, which involves the use of chemical additives, is likely to produce only a temporary improvement in electrode resistance, as the additives migrate and leach away over time, requiring their constant monitoring and replacement. Harmful environmental effects may also result and care must be taken to ensure that the chemicals used do not have an adverse effect upon the electrode material. Despite these disadvantages, however, soil treatment may prove to be the most economic solution for some temporary electrical installations in areas having high soil resistivity.

4.5 External earth fault loop impedance (Z_e)

The external earth fault loop impedance (Z_e) is the earth fault loop impedance of that part of the system that is external to the installation. It is one of the supply characteristics that must be determined by calculation, measurement or enquiry as required by BS 7671.

In almost every electrical installation, protection against electric shock by indirect contact is provided by EEBAD (earthed equipotential bonding and automatic disconnection of supply). Without a sufficiently low earth fault loop impedance, the earth fault current flowing under earth fault conditions will be insufficient to cause the protective device to operate and disconnect the faulty circuit within the maximum time permitted.

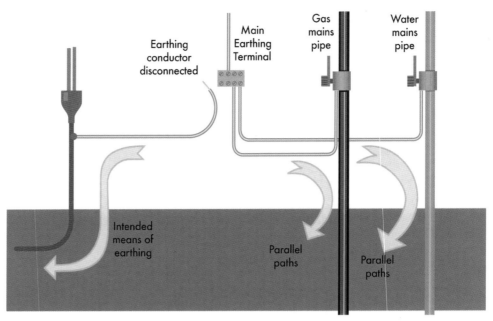

Figure 4.8 Parallel earth paths

Where earth fault loop impedance is to be determined by measurement, it is measured by testing at the origin of the installation. The purpose of the test is twofold:

1 To obtain an ohmic value for the earth fault loop impedance (Z_e)
2 To demonstrate that the intended means of earthing is present (normally the means of earthing is provided by the electricity distributor).

Prior to testing, the earthing conductor is disconnected from the main earthing terminal to remove parallel paths such as via the main equipotential bonding conductors, that might render the test result meaningless. A parallel earth path, via the main bonding, could hide the fact that the intended means of earthing is defective or non-existent.

Test instruments

Earth fault loop impedance instruments should only be used for the specific test application for which they were designed. Typically, an earth fault loop impedance measuring instrument determines impedance by charging a capacitor on a first half cycle and discharging that capacitance on the second half of the cycle with a resistance in series. The charge remaining is used to calculate the loop impedance.

Some of these instruments may also be used to measure fault level. This is the single-phase-to-earth fault level. For three-phase installations an approximation of the fault level can be obtained by doubling the single-phase-to-earth reading.

Because of the nature of the measurement, loop impedance testers may give misleading readings if used for any tests other than a standard loop test, or earth fault current measurement.

Most earth loop impedance testers are designed to inject an a.c. current to earth of about 20 A. This is sufficient to trip any RCD. To overcome this problem, earth loop impedance testers are available which use a d.c. current to desensitise the RCD during the test. This type of tester, however, only works on RCDs that are sensitive to a.c. faults alone. Type A RCDs (designed to the product standards BS EN 61008 and BS EN 61009) will trip upon detection of the d.c. desensitising current.

Earth loop impedance figures for installations which contain RCDs sensitive to both a.c. and d.c. fault currents (i.e. type A devices) should be determined either by calculation

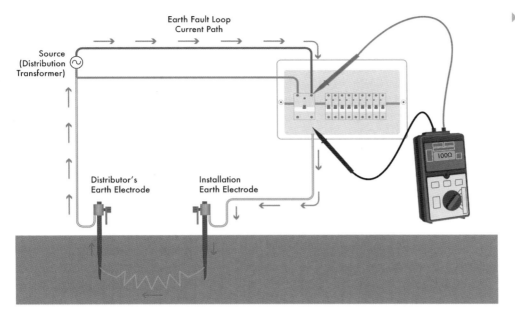

▶ **Figure 4.9** Measuring Z_e

or by using a tester having a test current below the device trip threshold. Alternatively, test methods can be used that will not trip the RCD. One such method is to measure the earth fault loop impedance on the supply side of the RCD and add this to the value of the combined resistance (R_1+R_2) on the load side of the RCD. This method also checks the continuity of the protective conductor, but it can only be used in an 'all insulated' installation.

Note (R_1+R_2) is the sum of the resistances of the phase conductor (R_1) and the circuit protective conductor (R_2) on the load side of the RCD.

Procedure for measuring Z_e

1 Securely isolate the installation by switching off the main switch and locking it in the open position.
2 Check that the test instrument and its probes and connections are in serviceable condition.
3 Check the test instrument using a proving instrument, use the test instrument to prove all conductors are dead and recheck the test instrument with the proving instrument.
4 Remove parallel paths by means such as disconnecting the earthing conductor from the main earthing terminal.
5 Make and record the measurement.
6 Reconnect the earthing conductor, main bonding conductors and circuit protective conductors that were disconnected in step 4.
7 Reenergise the installation.

Results

Where the distributor has provided a means of earthing, the maximum values measured for Z_e will be, typically, depending on the particular electricity distributor:

▶ TN-S supply   0.8 Ω
▶ TN-C-S supply   0.35 Ω

In a TT system, where the means of earthing for the installation is provided by an earth electrode such as a rod or a plate, the measured value of Z_e is likely to be much higher.

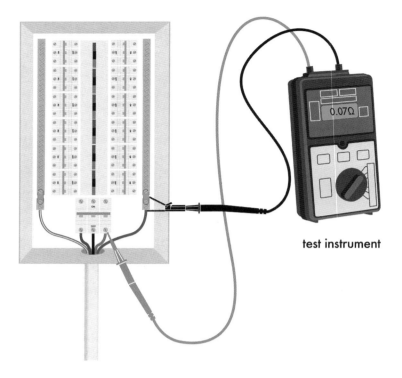

▶ **Figure 4.10**
Prospective fault current
testing on a three-phase
installation

test instrument

4.6 Prospective fault current

With the power on, the maximum value of the prospective short-circuit current (PSCC) can be obtained by direct measurement between live conductors close to the origin of the installation as shown. For three-phase supplies, the maximum balanced prospective short-circuit current level will be, as a rule of thumb, approximately twice the single-phase value. This figure errs on the side of safety.

These values then should be checked with the breaking capacity of the protective device to ensure that the breaking capacity is greater than the measured value of prospective fault current.

The prospective earth fault current (PEFC) may be measured with the same instrument by connecting between the phase and protective conductors.

4.7 Functional testing including the operation of residual current devices

All equipment in the electrical installation should be functionally tested.

Each RCD in an electrical installation that provides supplementary protection against direct contact should be functionally tested to verify its correct operation by a test instrument. This test must be independent of any facility (test button) incorporated in the device.

Although the fault may be simulated by a simple test, reliable operation of the RCD is best ensured by using an RCD tester and applying the range of tests described in Tables 4.2 and 4.3.

Test results are recorded on the test result schedule.

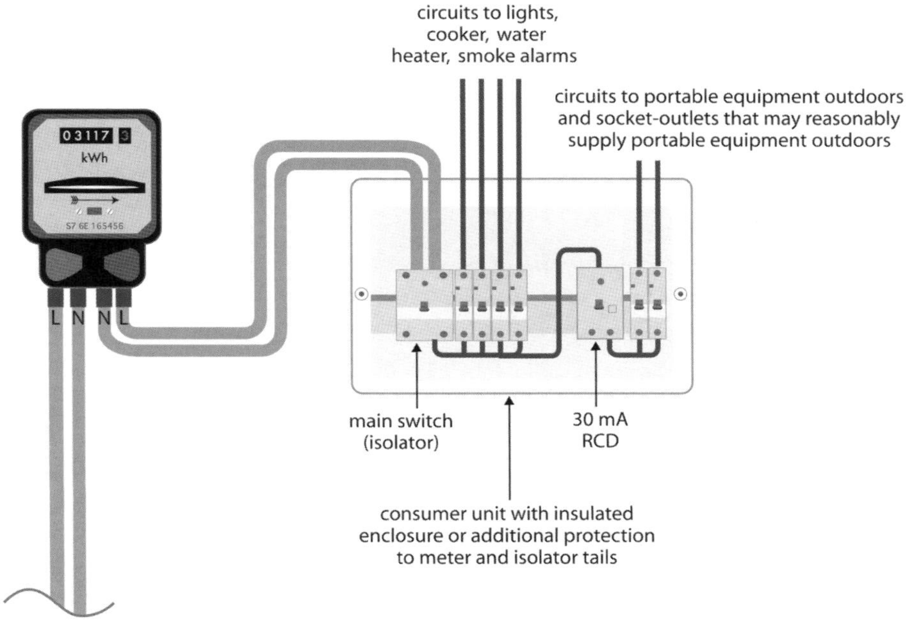

Figure 4.11 The 30 mA RCD in the split board must be tested

The tests are made on the load side of the RCD, as near as practicable to its point of installation, and between the phase conductor of the protected circuit and the associated circuit protective conductor. The load supplied should be disconnected during the tests.

General purpose RCDs to BS 4293 and RCD protected socket-outlets to BS 7288	
With a leakage current flowing equivalent to 50% of the rated tripping current	the device should not operate
With a leakage current flowing equivalent to 100% of the rated tripping current of the RCD	the device should operate in less than 200 ms
Where the RCD incorporates an intentional time delay	the device should trip within a time range from 50% of the rated time delay plus 200 ms to 100% of the rated time delay plus 200 ms

Table 4.2 General purpose RCDs to BS 4293 and RCD protected socket-outlets to BS 7288

General purpose RCDs to BS EN 61008 or RCBOs to BS EN 61009	
With a leakage current flowing equivalent to 50% of the rated tripping current of the RCD	the device should not operate
With a leakage current flowing equivalent to 100% of the rated tripping current of the RCD	the device should operate in less than 300 ms
	unless it is of 'Type S' (or selective) which incorporates an intentional time delay; in this case, it should operate within a time range from 130 ms to 500 ms

Table 4.3 General purpose RCDs to BS EN 61008 or RCBOs to BS EN 61009

© The Institution of Engineering and Technology

Additional requirement for supplementary protection

Where an RCD or RCBO with a rated residual operating current $I_{\Delta n}$ not exceeding 30 mA is used to provide supplementary protection against direct contact, with a test current of $5I_{\Delta n}$ the device should open in less than 40 ms. The maximum test time must not be longer than 40 ms, unless the protective conductor potential rises by less than 50 V (the instrument supplier will advise on compliance).

Integral test device

An integral test device is incorporated in each RCD. This device enables the electrical and mechanical parts of the RCD to be verified, by pressing the button marked 'T' or 'Test'. Operation of the integral test device does not provide a means of checking: (a) the continuity of the earthing conductor or the associated circuit protective conductors, or (b) any earth electrode or other means of earthing, or (c) any other part of the associated installation earthing. The test button will only operate the RCD if the RCD is energised. The notice to test RCDs quarterly (by pressing the test button) must be fixed in a visible position.

Test failure

If a test shows a failure to comply, the installation fault must be corrected. The test must then be repeated, as must any earlier test that could have been influenced by the failure.

▶ **Figure 4.12** RCD testing notice

The installation, or part of it, is protected by a device which automatically switches off the supply if an earth fault develops. Test quarterly by pressing the button marked 'T' or 'Test'. The device should switch off the supply and should then be switched on to restore the supply. If the device does not switch off the supply when the button is pressed, seek expert advice.

Lighting maintenance

5

5.1 Introduction

A lighting installation should be maintained to keep its visual performance within the design limits. The designer will have selected a certain illumination level for the particular activity and presumed a frequency of lamp replacement and a frequency of cleaning. The frequency of lamp replacement and cleaning will be appropriate to the environment, including accessibility and the type of luminaire (light fitting). When assessing maintenance requirements the first step is to seek information on the initial design assumptions. Maintaining a lighting installation as intended will ensure the efficiency of the installation is not degraded. Reduced maintenance may result in reduced operational performance and, in extremes, lead to danger. Maintaining a lighting installation in good order is also important for maintenance of staff morale and provision of a good impression to customers – flickering, failed and discoloured lamps may discourage staff and turn customers away.

There are two aspects to luminaire maintenance:

1. luminaire cleaning
2. lamp replacement.

5.2 Luminaire cleaning

5.2.1 Frequency
The frequency of luminaire cleaning depends upon three factors:

1. the type of luminaire and its inherent ability to maintain light output over a period for a given environment
2. the environment (whether it is clean, normal or dirty)
3. the permitted (or assumed for design purposes) reduction in light output before luminaire cleaning is required.

The reduction in light output as a proportion of initial light output is called the luminaire maintenance factor (LMF) and is given in Table 5.1.

The following procedure can be followed for determining the intervals between luminaire cleaning:

1. determine luminaire maintenance category from Table 5.2
2. assess the environment (see Table 5.3)
3. identify the luminaire maintenance factor.

Reference to Table 5.1 will then give the permissible period between luminaire cleaning.

▶ **Figure 5.1** GLS (general lighting service) lamp

Luminaire maintenance factor (LMF) (light output as a proportion of initial light output)[1]																		
Elapsed time between cleanings in years	0.5 years			1.0 year			1.5 years			2.0 years			2.5 years			3.0 years		
Environment	clean	normal	dirty	clean	normal	dirty	clean	normal	dirty	clean	normal	dirty	clean	normal	dirty	clean	normal	dirty
Luminaire category																		
A	.95	.92	.88	.93	.89	.83	.91	.87	.80	.89	.84	.78	.87	.82	.75	.85	.79	.73
B	.95	.91	.89	.90	.86	.83	.87	.83	.79	.84	.80	.75	.82	.76	.71	.79	.74	.68
C	.93	.89	.83	.89	.81	.72	.84	.74	.64	.80	.69	.59	.77	.64	.54	.74	.61	.52
D	.92	.87	.83	.88	.82	.77	.85	.79	.73	.83	.77	.71	.81	.75	.68	.79	.73	.65
E	.96	.93	.91	.94	.90	.86	.92	.88	.83	.91	.86	.81	.90	.85	.80	.90	.84	.79
F	.92	.89	.85	.86	.81	.74	.81	.73	.65	.77	.66	.57	.73	.60	.51	.70	.55	.45

▶ **Table 5.1** Luminaire maintenance factor (LMF) (light output as a proportion of initial light output)[1]

1 Luminaire maintenance factor (LMF) is given by: $\text{LMF} = \dfrac{\text{light output after a specified time}}{\text{initial light output}}$

The designer of the lighting installation will have selected a luminaire maintenance factor (LMF); however, in the absence of such information a factor of 0.8 is usually assumed.

▶ **Table 5.2** Luminaire maintenance categories

Category	Representation	Description
A		the light source is in free air there are no reflector surfaces, diffusers, or covers to be contaminated
A		GLS lamp
B		white painted metal reflectors with slots in the top to create air currents to keep the reflectors clean
B		high bay luminaires
B		PAR 38 reflector lamp
B		GLS lamp with reflector
B		compact fluorescent with reflector

continues

Category	Representation	Description
C		white painted or aluminium reflectors without top slots for air currents
C		low bay luminaires with reflectors
D		recessed luminaires
E		surface modular luminaires
E		general diffusing luminaires
F		uplighters

▶ **Table 5.2** *continued*

Environment	Examples
Clean	Offices, shops, laboratories
Normal	Light industrial, outdoor
Dirty	Contamination by smoke, dust etc. e.g. foundries, rubber processing

▶ **Table 5.3** The environment

Assuming a luminaire maintenance factor of 0.8, cleaning intervals have been determined in Table 5.4 for selected luminaires and maintenance types.

Luminaire maintenance category	Typical luminaire type	Environment		
		clean	normal	dirty
A	fluorescent no reflector	3 yrs	2	1
B	high bay	3	2	1
C	fluorescent aluminium reflector	2	1	$\frac{1}{2}$
D	recessed	2	1	$\frac{1}{2}$
E	general	3	3	2
F	uplighter	1	1	$\frac{1}{2}$ yr

▶ **Table 5.4** Luminaire cleaning intervals (years; LMF = 0.8)

Figure 5.2 PAR 38 lamp

5.2.2 Wall and ceiling cleaning

Room lighting levels, as well as depending upon the cleanliness of the luminaires, also depend on the cleanliness of the room, particularly ceilings and walls. This factor is called the room surface maintenance factor (RSMF). The designer will have assumed a factor for this and may well have assumed a cleaning time for the walls and ceiling. Again, the intervals between the cleaning of floors and ceilings will depend upon the environment and to a lesser degree the nature of the lighting, i.e. whether it is direct or indirect, and on the size of the room. Indirect lighting reflecting from a ceiling is very dependent upon the cleanliness of the ceiling or the surface from which the lighting is being reflected, and dirty walls will have a lesser impact on a large room than on a small room.

5.2.3 Precautions when cleaning luminaires

The following are precautions that should be taken when cleaning luminaires:

▶ Ensure the luminaire is switched off and has cooled down.
▶ Extreme caution should be exercised when cleaning all surfaces of luminaires. Some surfaces are very susceptible to abrasion; for example, polished (unanodised) aluminium is easily scratched, as are some plastics.
▶ The maintainer should experiment on a small test area with the proposed cleaning method before starting.
▶ Care is required in handling plastics, as they tend to become brittle with age. Depending on the environment or light source, some plastics may also turn yellow. There is no successful way of cleaning when this happens, and replacement must be considered.

5.2.4 Cleaning methods

Aluminium reflectors should be washed with a warm, soapy solution and rinsed thoroughly before being air dried. Plastic opal or prismatic lenses should be cleaned with a damp cloth (using non-ionic detergent and water) and treated with antistatic polish or spray and allowed to dry. Vitreous enamel, stove enamel and glass optics should be wiped with a damp cloth using a light concentration of detergent in water.

For best results, louvre (rectangular and square cell) optics should be cleaned by removing the louvre and dipping it in a warm water solution.

Specular finished (particularly plastic) louvres are very difficult to clean, and appearance deteriorates over the years. Therefore, they should be used only where air quality is very clean, such as new office buildings, banks etc.

5.2.5 Cleaning agents

Choice of cleaning materials and methods is determined by the type of dirt to be removed and the type of material to be cleaned. For plastic materials a final treatment with an antistatic substance is recommended.

General cleaning – The first and most commonly used is a dry chemical detergent with additives in different concentration levels. It is an advantage to use compounds that require no rinsing after the wash.

Heavy duty cleaning of oil concentrations (e.g. in auto garages, oily factories etc.) – The second type of cleaner is a heavy duty liquid cleaner which may contain detergents, solvents and abrasives. It is particularly useful for the removal of oily dirt, but must be tested to ensure that it does not damage materials or leave deposits.

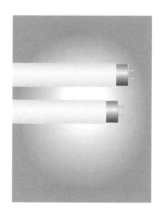

Figure 5.3 Fluorescent tubes

Excessively oily industrial conditions – In some heavy oily applications, the use of a high pressure steam cleaner is practicable provided the system has been designed with this cleaning technique in mind.

5.3 Lamp replacement

5.3.1 General
Three factors are particularly important when considering the frequency of lamp replacement. These are:

1 the lamp survival factor, LSF (the proportion of lamps still working after a specified burning time)
2 the lamp lumen maintenance factor, LLMF (the proportion of the initial light output being maintained after a specified burning time due to deterioration (ageing) of the lamp)
3 the cleaning frequency.

▶ **Figure 5.4** Compact fluorescent lamp (CFL)

5.3.2 Lamp survival factor (LSF)
Lamps fail in service, some more frequently than others. This is particularly important with incandescent lamps, as can be seen from Table 5.5. For other types of lamp, failure is unlikely until the burning time has exceeded some thousands of hours. What is likely to precipitate lamp change is the reduction in the initial light output and the environmental effects of failed or flickering lamps. Although a failed or flickering lamp may not make much difference to the illumination level, it is both disturbing and distracting to staff and customers.

The maintenance engineer has to decide what is an acceptable lamp survival factor not only from a lumen output point of view but also from appearance. One in twenty lamp failures may be just acceptable in a light industrial workshop but would probably be unacceptable in an office environment. The longer the intervals between group lamp replacement the greater are the costs associated with individual lamp replacements. The maintenance engineer cannot generally refuse a request to replace a failed lamp in an office environment.

5.3.3 Lamp lumen maintenance factor (LLMF)
The lamp lumen maintenance factor (LLMF) can be confused with the luminaire maintenance factor (LMF). The LLMF is concerned with reductions in light output of the lamp (not luminaire) as a result of ageing. Cleaning does not change the LLMF. The luminaire maintenance factor (LMF) is concerned with dirt on the luminaire.

As a lamp burns, it discolours and its light output reduces. The LLMF is a measure of this and is given by:

$$\text{LLMF} = \frac{\text{light output after a specified burning time}}{\text{initial light output}}$$

5.3.4 Cleaning frequency
It is cost effective to replace lamps when carrying out routine cleaning of the luminaires. As a result, it is almost always sensible to arrange lamp replacement during routine cleaning, but perhaps not at every routine cleaning.

▶ **Figure 5.5** Bulkhead luminaire

Lamp type	Incandescent GLS	Fluorescent multi and tri-phosphor	Fluorescent halophosphate	Mercury	Metal halide	High pressure sodium	High pressure sodium-improved colour
Factors	LLMF / LSF	LLMF / LSF	LLMF / LSF	LLMF / LSF	LLMF / LSF	LLMF / LSF	LLMF / LSF
0.1	1.00 / 1.00	1.00 / 1.00	1.00 / 1.00	1.00 / 1.00	1.00 / 1.00	1.00 / 1.00	1.00 / 1.00
0.5	.97 / .98	.98 / 1.00	.97 / 1.00	.99 / 1.00	.96 / 1.00	1.00 / 1.00	.99 / 1.00
1.0	.93 / .50	.96 / 1.00	.94 / 1.00	.97 / 1.00	.93 / .97	.98 / 1.00	.97 / 1.00
1.5	.89 / .30	.95 / 1.00	.91 / 1.00	.95 / 1.00	.90 / .96	.97 / 1.00	.95 / .99
2.0		.94 / 1.00	.89 / 1.00	.93 / .99	.87 / .95	.96 / .99	.94 / .98
4.0		.91 / 1.00	.83 / 1.00	.87 / .98	.78 / .93	.93 / .98	.89 / .96
6.0		.87 / .99	.80 / .99	.80 / .97	.72 / .91	.91 / .96	.84 / .90
8.0		.86 / .95	.78 / .95	.76 / .95	.69 / .87	.89 / .94	.81 / .79
10.0		.85 / .85	.76 / .85	.72 / .92	.66 / .83	.88 / .92	.79 / .65
12.0		.84 / .75	.74 / .75	.68 / .88	.63 / .77	.87 / .89	.78 / .50
14.0		.83 / .64	.72 / .64	.64 / .84	.60 / .70	.86 / .85	
16.0		.81 / .50	.70 / .50	.61 / .80	.56 / .60	.85 / .80	
18.0				.58 / .75	.52 / .50	.83 / .75	
20.0				.55 / .68		.82 / .69	
22.0				.53 / .59		.81 / .60	
24.0				.52 / .50		.80 / .50	

Burning (thousand hours)

▶ **Table 5.5** Lamp lumen maintenance factors and lamp survival factors

5.3.5 Example calculation of lamp replacement

Having decided upon:

▸ the lamp survival factor (LSF), and
▸ the lamp lumen maintenance factor (LLMF)

the frequency of lamp replacement can be calculated.

Consider a fluorescent lamp installation (multiphosphor) and assume the LSF must be 0.95 or greater and the LLMF 0.8. From Table 5.6:

1 LSF 0.95, therefore lamps must be replaced before 8000 h.
2 LLMF 0.8, therefore lamps must be replaced before 16 000 h.

Therefore replace lamps at lowest of 1 and 2, that is 8000 h.

The lamp burning hours need to be estimated. In most commercial premises little natural light is available and hours of occupancy are a very good guide to hours of burning.

Assuming a nine-hour day, six days a week for 50 weeks:

$$9 \times 6 \times 50 = 2700 \text{ h}$$

burning per annum.

In the example above, the lamps would need replacing after

$$\frac{8000}{2700} = 3 \text{ years (approx.)}$$

After an estimation of the cleaning intervals as described in Section 5.2, decisions can be made on the frequency of the combined activities of cleaning and lamp replacement. In Table 5.6 examples have been prepared that include the calculation of cleaning intervals, and lamp replacement frequency.

▸ **Figure 5.6** Sodium or mercury discharge luminaire

Table 5.5 notes:
LLMF − (lamp lumen maintenance factor): proportion of the initial light output emitted after a specified burning time
LSF − (lamp survival factor): proportion of lamps surviving after a specified burning time

Location		Office	Light industry	Street lighting
Luminaire type		Fluorescent multi phosphor B	Fluorescent multi phosphor B	Sodium E
Occupancy	Description	6 days per week	2 shifts, 6 days per week	all night half night
	Days/year	300	300	365
	Hours/day	9	16	
Lamp burning time hrs/year		2 700	4 800	4 000 2 000
Adopted lamp survival factor LSF		0.95	0.9	0.9 0.9
Burning hours to LLMF from Table 5.5		8 000	9 000	10 000 10 000
Years to LSF		2.96	1.88	2.5 5
Adopted lamp lumen maintenance factor LLMF		0.80	0.85	0.80 0.80
Burning hours to LLMF from Table 5.5		16 000	10 000	24 000 24 000
Years to LLMF		5.9	2.0	6 12
Selected luminaire maintenance factor		0.8	0.8	0.8 0.8
Environment		clean	normal	normal normal
Time between cleaning from Table 5.1		3.0	2.0	3+ 3+
Selection lamp	Change	3	3	3[1] 6[1]
	Clean	3	2	3 3

▷ **Table 5.6** Examples of selection of lamp change and luminaire cleaning frequencies

Note:
1 with failed lamps replaced as necessary

5.4 Lamp disposal

Chapter 12 of this publication discusses lamps as special waste.

5.4.1 Accidental breakage of a lamp
The accidental breakage of a lamp gives rise to hazards associated with:

▶ flying glass
▶ fire and combustive explosion
▶ toxic and corrosive substances.

There is an ongoing hazard from flying glass and the handling of glass fragments. In the event of an accidental breakage of a lamp, normal good housekeeping is required, care being necessary to prevent injury from broken glass. For fluorescent lamps the generation and inhalation of airborne dust should be avoided when cleaning up; for low-pressure sodium lamps avoid skin and eye contamination with debris and prevent exposure to moisture.

5.4.2 Hazards associated with particular lamps
Low pressure sodium lamps
Low pressure sodium lamps (SOX) contain sodium metal, which reacts with water. Hazards to be considered are the potentially corrosive sodium hydroxide solution and the extremely flammable and explosive hydrogen gas, which result from reacting sodium with water.

Discharge lamps
Discharge lamps contain small quantities of toxic materials, including lead and mercury, which may be released as dust or vapour.

Linear fluorescent lamps
Some lamps, especially fluorescent lamps, may release powders when broken. The powder may be contaminated with mercury, and the inhalation of any dust must be avoided.

High intensity discharge lamps
High intensity discharge lamps have outer and inner envelopes. The inner arc tubes are fairly strong and contain substances such as small quantities of mercury, sodium and other metals. High pressure sodium lamps also contain sodium, although less than SOX, and the sodium is contained in the inner quartz tube.

▶ **Figure 5.7** Halogen downlighter

▶ **Figure 5.8** Fluorescent luminaire fitted in a suspended ceiling

In-service inspection and testing of electrical equipment

6

6.1 The need to inspect and test

As discussed in Chapter 3, the *Electricity at Work Regulations* require all electrical systems to be maintained in a safe condition. The definition of a system includes all the equipment and apparatus of the fixed electrical installation of the building and equipment and appliances supplied from the fixed electrical installation. There is a wide range of portable appliance testers available to test appliances, though the expression 'portable' is a little misleading as the equipment may be portable, movable, hand-held, stationary, fixed or suitable only for building-in. However, all equipment requires inspection and testing and this can generally be effected by a portable appliance tester or by the use of an insulation and continuity tester.

The Institution of Electrical Engineers publishes a *Code of practice for in-service inspection and testing of electrical equipment*. This code has been prepared both for administrators with responsibility for electrical maintenance who may not necessarily have the technical knowledge to perform the necessary inspection and testing, and for the staff who carry out the inspection and testing. It is comprehensive guidance prepared in cooperation with a range of trade organisations and government departments including the Health and Safety Executive.

6.2 Types of test

Electrical equipment is inspected and tested at the following stages in its life:

- type testing to a British Standard – carried out as part of the approval procedure for the appliance
- routine end-of-line testing – carried out on each appliance by the manufacturer before sale
- in-service inspection and testing
- testing after repair.

In-service inspection and testing is discussed briefly in this chapter and more fully in the IEE's *Code of practice for in-service inspection and testing of electrical equipment.*

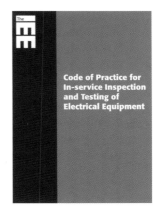

Figure 6.1 Appliances may be portable, movable, hand-held, stationary, fixed or suitable for building-in

Figure 6.2 The IEE's *Code of practice for in-service inspection and testing of electrical equipment*

6.3 In-service inspection and testing

In-service inspection and testing falls into three categories:

1 *User checks* Inspections carried out by the user. All faults found are reported and logged but no record is required if no fault is found.
2 *Formal visual inspections* Inspections without tests, the results of which are satisfactory or not satisfactory and are recorded.
3 *Combined inspection and test* Generally performed by an electrical operative and the results are formally recorded.

6.4 Frequency of inspection and testing

The *Electricity at Work Regulations* require equipment to be maintained so as to prevent danger. The Regulations do not specify the frequency of inspection and testing of equipment. The responsible person, normally the user, must determine how frequently equipment needs checking, inspecting and testing. Factors that would influence the frequency of inspection and testing include:

The environment Equipment installed in a controlled environment such as an office will suffer less wear, tear and damage than equipment in an arduous environment such as on a construction site.

Typical use The care exercised by the users and the likelihood of any damage to the equipment being reported.

Equipment construction The robustness and suitability of the equipment will affect the frequency of inspection and testing.

▶ **Figure 6.3** Most washing machines are Class I equipment

6.5 Equipment construction

Equipment construction is divided into three classes, Class I, Class II and Class III. Most equipment encountered will be Class I or Class II.

6.5.1 Class I equipment
Class I equipment is equipment in which protection against electric shock does not rely solely on the basic insulation of the equipment, but also includes a means for connection of exposed-conductive-parts to a protective conductor (earth wire) in the fixed wiring of the electrical installation. In layman's terms, Class I equipment requires a connection with Earth and this connection with Earth must be maintained to ensure the safety of the equipment.

6.5.2 Class II equipment
Class II equipment is equipment in which protection against electric shock does not rely on basic insulation only but in which additional insulation such as supplementary insulation is provided. There is no provision for the connection of any exposed metalwork of the equipment to a protective conductor and no reliance upon protective devices in the fixed installation of the building.

Certain information technology Class II equipment may have a functional earth connection, necessary for the proper functioning of the equipment but which does not play a part in the electrical safety of the equipment.

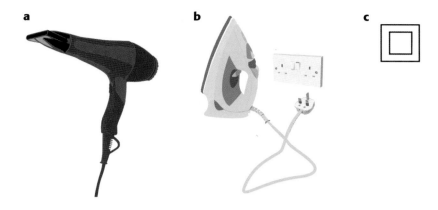

a b c

Figure 6.4 Class II equipment.
a hairdryers are generally Class II equipment
b many irons are Class II equipment
c symbol for Class II equipment

6.5.3 Class III equipment

Class III equipment is equipment in which protection against electric shock relies on a supply of electricity from a Separated Extra-Low Voltage (SELV) source such as an isolating transformer to BS 3535 or BS EN 60742.

a

b

Figure 6.5 Class III equipment.
a symbol for Class III equipment
b symbol for safety isolating transformer

6.6 Recommended initial frequencies of inspection and testing

Table 6.1 provides guidance on suggested initial frequencies of inspection and testing of equipment. References in the notes below the Table are references to clauses in the IEE Code of Practice.

User checks are not normally recorded unless a fault is found. The equipment will then be removed from service and entered in the faulty equipment register.

The formal visual inspection may form part of the combined inspection and tests when they coincide, and must be recorded (see Section 6.3).

If the Class of equipment is not known, it should be tested as Class I.

6.7 Review of frequency of inspection and testing

The intervals between user checks, visual inspections and combined inspection and testing must be kept under review, particularly until patterns of failure and damage are determined. Users must be encouraged to report faults found in their more frequent checks. After some experience has been gained, it may be possible to extend the intervals between formal visual inspections and combined inspection and tests.

6.8 Records

The *Memorandum of Guidance on the Electricity at Work Regulations* prepared by the Health and Safety Executive advises that records of maintenance including test results should be kept throughout the working life of equipment. Without such records, management cannot review the frequency of inspection and testing and determine whether the intervals need to be reduced or may be increased. To be confident that all equipment has been inspected and tested, all equipment should be permanently and uniquely marked or labelled and records kept.

▶ **Table 6.1** Initial frequency of inspection and testing of equipment

Type of premises	Type of equipment[1]	User checks[2]	Class I equipment		Class II equipment[4]	
			Formal visual inspection[3]	Combined inspection and testing[5]	Formal visual inspection[3]	Combined inspection and testing[5]
Construction sites 110 V equipment	S	none	1 month	3 months	1 month	3 months
	IT	none	1 month	3 months	1 month	3 months
	M[8]	weekly	1 month	3 months	1 month	3 months
	P[8]	weekly	1 month	3 months	1 month	3 months
	H[8]	weekly	1 month	3 months	1 month	3 months
Industrial including commercial kitchens	S	weekly	none	12 months	none	12 months
	IT	weekly	none	12 months	none	12 months
	M	before use	1 month	12 months	3 months	12 months
	P	before use	1 month	6 months	3 months	6 months
	H	before use	1 month	6 months	3 months	6 months
Equipment used by the public	S	6+7	monthly	12 months	3 months	12 months
	IT	6+7	monthly	12 months	3 months	12 months
	M	6+7	weekly	6 months	1 month	12 months
	P	6+7	weekly	6 months	1 month	12 months
	H	6+7	weekly	6 months	1 month	12 months
Schools; user checks to be performed by supervisor, teacher or member of staff	S	weekly[7]	none	12 months	12 months	48 months
	IT	weekly[7]	none	12 months	12 months	48 months
	M	weekly[7]	4 months	12 months	4 month	48 months
	P	weekly[7]	4 months	12 months	4 month	48 months
	H	before use	4 months	12 months	4 month	48 months
Hotels	S	none	24 months	48 months	24 months	none
	IT	none	24 months	48 months	24 months	none
	M	weekly	12 months	24 months	24 months	none
	P	weekly	12 months	24 months	24 months	none
	H	before use	6 months	12 months	6 months	none
Offices and shops	S	none	24 months	48 months	24 months	none
	IT	none	24 months	48 months	24 months	none
	M	weekly	12 months	24 months	24 months	none
	P	weekly	12 months	24 months	24 months	none
	H	before use	6 months	12 months	6 months	none

Notes:

1 S = stationary equipment, IT = information technology equipment, M = movable equipment, P = portable equipment, H = hand-held equipment
2 user checks need not be recorded unless a fault is found
3 the formal visual inspection may form part of the combined inspection and tests when they coincide, and must be recorded
4 if the class of equipment is not known, it must be tested as Class I
5 the results of combined inspections and tests must be recorded
6 for equipment such as children's rides a daily check may be necessary
7 by a supervisor or teacher or member of staff
8 110 V earthed centre-tapped supply (or 110 V three-phase supply, 63.5 V to earth).
230 V portable or hand-held equipment on a construction site must be supplied via a residual current device with a residual operating current not exceeding 30 mA and inspections and tests carried out more frequently.

The following records are recommended:

i   a register of all equipment
ii a record of formal inspection and tests
iii a record of faulty equipment
iv a register of repairs.

Examples of the four records are available in the *Code of practice for in-service inspection and testing of electrical equipment*.

6.9 Inspection and testing

A user check is a common-sense visual inspection carried out by the user of the equipment. Formal visual inspections and combined inspections and tests may be carried out in-house by suitably competent and trained staff. The IEE *Code of practice for in-service inspection and testing of electrical equipment* provides guidance on the experience and training required to carry out both formal visual inspections and combined inspections and tests.

6.10 User checks

The user check is an important safety check. Many faults can be identified by a straightforward visual inspection. Detailed electrical knowledge is not necessary. The user is the person most familiar with the equipment and may be in the best position to know if it is in a safe condition and working properly. No record need be made of a user check unless some aspect of the inspection is unsatisfactory.

The user check should proceed as follows:

a The user should consider whether there is any known fault with the equipment and whether it works properly
b The user then disconnects the equipment, if appropriate (as described in Section 14 of the Code of Practice)
c The user then inspects the equipment looking at:
  i The flexible cable (where fitted) which should:
    ▸ be in good condition
    ▸ be free from cuts, fraying and damage
    ▸ run in a place where it will not be damaged
    ▸ not run where it might be a trip hazard
    ▸ not run under carpets, for example, where it might overheat
    ▸ be neither too long nor too short
    ▸ not have any joints.
  ii The plug (where fitted):
    ▸ the plug must be wired correctly and the screws must be tight
    ▸ the correct fuse must be fitted
    ▸ the flexible cable should be secure in its clamp
    ▸ there should be no sign of overheating
    ▸ there should be no sign of cracks or damage.

▸ **Figure 6.6** Plug removed from socket-outlet

▸ **Figure 6.7** Flex with kink, damage and taped joint

▸ **Figure 6.8** The plug MUST be wired correctly

▸ **Figure 6.9** Cable pulled out of plug

▶ **Figure 6.10** Cracked and overheated socket-outlet

iii The socket-outlet or flex outlet:
  ▶ the outlet must be in good condition
  ▶ there should be no sign of overheating
  ▶ there should be no sign of cracks or damage.
iv The equipment:
  ▶ the equipment should work correctly
  ▶ it should switch on and off properly
  ▶ it should be undamaged and free from cracks
  ▶ it should be able to be used safely.
v The equipment should be suitable for its environment.
vi The equipment should be suitable for the work it is required to carry out.
vii There should not be excessive use of extension leads, cube adapters or multiway adapters.
d The user should take action if any faults or damage are observed:
  i Faulty equipment must be switched off and unplugged from the supply.
  ii A faulty item needs to be labelled to emphasise that it must not be used (if possible, move it to a place of safe keeping).
  iii The fault should be reported to the responsible person.
  iv The faulty item needs to be entered in the faulty equipment log.
  v The item needs to be repaired.

If equipment is found to be damaged or faulty on inspection or test, an assessment must be made by a responsible person as to the suitability of the equipment for its use and its location. Frequent inspections and tests will not prevent damage occurring if the equipment is unsuitable. Replacement by suitable equipment is required.

6.11 Formal visual inspections and combined inspections and tests

Formal visual inspections and combined inspections and tests should be carried out by a competent person following the procedures set out in the IEE *Code of practice for in-service inspection and testing of electrical equipment.*

▶ **Figure 6.11** Faulty appliance

Emergency lighting   7

7.1 What is emergency lighting?

Emergency lighting is a primary life safety system provided to assist occupants to evacuate in the case of an emergency and, if necessary, to permit certain tasks, such as a controlled shutdown, to be performed in safety.

Emergency lighting includes standby lighting and emergency escape lighting which is subdivided into escape route lighting, open area lighting and high risk task area lighting.

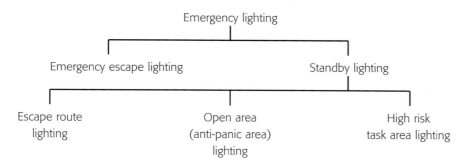

▶ **Figure 7.1** Emergency lighting types

Emergency escape lighting – that part of emergency lighting provided to enable safe exit in the event of failure of the normal supply.

Standby lighting – that part of emergency lighting provided to enable normal activities to continue in the event of failure of the normal mains supply.

Escape route lighting – that part of emergency lighting provided to enable safe exit for building occupants by providing appropriate visual conditions and direction finding on escape routes and in special areas/locations, and to ensure that fire fighting and safety equipment can be readily located and used.

Open area (or anti-panic) area lighting – that part of emergency escape lighting provided to reduce the likelihood of panic and to enable safe movement of occupants towards escape routes by providing appropriate visual conditions and direction finding.

High risk task area lighting – that part of emergency lighting provided to ensure the safety of people involved in a potentially dangerous process or situation and to enable proper shutdown procedures to be carried out for the safety of other occupants of the premises.

Emergency lighting is provided to prevent a hazard in the event of the loss of supply to the normal lighting installation. One of the most important functions of the emergency lighting is to provide reassurance to occupants and to allow orderly and speedy evacuation of a building, should this be necessary. Although, in general, emergency lighting is considered to be escape lighting, all hazards that might arise as a result of

© The Institution of Engineering and Technology

loss of the normal lighting must be considered. Emergency lighting may be required to illuminate switchrooms and control rooms, to facilitate restoration of supplies or management of facilities to allow dangerous plant to be shut down.

Emergency lighting for areas where artificial lighting is required night and day has to be considered from basic principles. There may be a need to illuminate the area to prevent danger in the event of loss of electricity supply as well as a need to provide escape lighting.

7.1.1 Maintained and non-maintained

Two basic types of luminaire are specified – maintained and non-maintained:

Maintained emergency lighting (M) – a lighting system in which all emergency lighting lamps are in operation at all material times.

Non-maintained emergency lighting (NM) – a lighting system in which all emergency lighting lamps are in operation only when the supply to the normal lighting fails.

Self-contained emergency luminaires may be either maintained or non-maintained. Such luminaires contain all the elements of the emergency luminaire, including the battery, the lamp, the control unit and test and monitoring facilities.

7.1.2 Categories

Luminaires are categorised by being maintained or non-maintained and by the number of hours, the duration, for which they can maintain their light output to an acceptable level after supply failure.

For example, a non-maintained luminaire with a duration of two hours is given the designation NM/2 and a maintained luminaire with a duration of three hours is given the designation M/3.

In most cases evacuation of the building will not take longer than one hour and a duration of one hour will be sufficient. However, in certain conditions the local authority licence may allow a period of time for continued occupation after the failure of the normal lighting. In these situations, the minimum duration of the emergency lighting should be one hour after any such period of permitted occupation. Additionally, particularly in larger premises, emergency lighting will be necessary after the evacuation of the building has been substantially completed, for safety requirements such as searching of the premises to ensure that no persons have been left behind or to allow reoccupation of the premises in order to get people off the street and into a place of relative safety.

7.1.3 Emergency lighting categories

Emergency lighting categories are given in Table 7.1.

7.2 The need for emergency lighting

The basic risk to be assessed is that of fire and the means of escape in case of fire. A fire detection and alarm system may be required to alert people in the early stages of a fire and as normal lighting is likely to fail during a fire, emergency lighting should be provided to facilitate escape.

Emergency lighting can and does save lives.

Type of premises			Minimum category
Premises used as sleeping accommodation	hospitals, nursing homes, hotels, guest houses, clubs, colleges and schools	ten bedrooms or more	NM/3 or M/3
		small – not more than ten bedrooms or not more than one floor above or below ground level	NM/2 or M/2
Non-residential premises used for treatment or care	special schools, clinics and the like		NM/1
Non-residential premises used for recreation	theatres, concert halls, discotheques, exhibition halls, sports halls, public houses, restaurants	premises where there is facility to dim the normal lighting or for the consumption of alcohol	M/2
		More than 250 persons present	NM/2
		small – no more than 250 persons present	M/1 or NM/1
Non-residential used for recreation	ballrooms and dance halls cinemas (licensed under the *Cinemas Act 1985*) bingo premises (licensed under the *Gaming Act 1968* as amended) ten pin bowling premises		Not covered by BS 5266
Non-residential used for teaching, training and research	schools, colleges, technical institutes and laboratories		NM/1
Non-residential public premises	town halls, libraries, offices, shops, art galleries, museums	Premises where lighting may be dimmed	M/1
		Other	NM/1
Industrial premises used for manufacture, processing or storage of products	factories, workshops, warehouses and similar establishments		NM/1 or M/1
Multiple use of premises	premises falling into one or more of the above classes should be treated in accordance with the most stringent conditions unless the different uses are contained within separate fire compartments with independent escape routes		apply the most stringent conditions
Common access routes within multi-storey dwellings	buildings up to ten storeys		NM/1
	buildings greater than ten storeys		NM/3
Enclosed shopping malls	walkways and escape routes		M/2
	commercial premises situated off such escape routes		NM/1
Covered car parks	apply the same conditions as for escape routes in non-residential public premises		NM/1 or M/1
Sports stadia	refer to the Home Office's *Guide to safety at sports grounds*		

▶ **Table 7.1** Emergency lighting categories

The *Fire Precautions (Workplace) Regulations 1997* require that: 'Emergency routes and exits must be indicated by signs and emergency routes and exits requiring illumination shall be provided, where necessary, with emergency lighting of adequate intensity in the case of failure of their normal lighting'.

7.3 Escape lighting

Escape lighting must:

▶ indicate the escape routes
▶ illuminate these escape routes
▶ illuminate fire alarm call points and fire fighting equipment.

Escape lighting must be provided not only as a consequence of complete supply failure but also on local failure. For example, escape lighting must be available should a single lighting final circuit supplying luminaires in a stairwell stop working.

7.4 Design considerations for an emergency lighting system

Before designing an emergency lighting system the following information needs to be determined from the site drawings or from the specifier:

a The duration of the emergency lighting. Three-hour duration is required in places of entertainment and for sleeping risk and if evacuation is not immediate, or early reoccupation is likely to occur. One-hour duration may be acceptable, in some premises, if evacuation is immediate and reoccupation is delayed until the system has recharged.
b Emergency lighting should be of the maintained type in areas in which the normal lighting can be dimmed. In addition, emergency lighting must be of the maintained type in common areas within shopping malls where a build-up of smoke could reduce the effectiveness of normal lighting.
c Exit signs must be of the maintained type where the premises are used by people who are unfamiliar with its layout.
d Building plans need to be assembled showing the location of the fire alarm call point positions, the positions of fire fighting equipment and fire and safety signs.
e Emergency escape routes should be established, and potential hazards investigated.
f Open areas larger than 60 m² floor area should be identified.
g High risk task areas should be identified and normal lighting levels established.
h The need for external illumination outside final exit doors and on a route to a place of safety should be determined.
i Other areas that need illumination, although not part of the escape route, should be located, e.g. lifts, moving stairways and walkways, plant rooms and toilet accommodation over 8 m² gross area.
j If a central system is being used, the location of central battery units and cable runs should be established in areas of low fire risk.
k For non-maintained applications the area covered by the final circuit of the normal lighting has to be determined, as it must be monitored by the central system. Non-maintained self-contained luminaires must be fed from that final circuit.
l Standby lighting requirements should be established if activities need to continue during a failure of the normal lighting supply.

m The customer's preference and operating considerations should be ascertained, e.g. ceiling heights, mounting heights or wall mounting.

n If the incoming mains electrical supply to the premises is not in an area of low risk, then staff may be exposed to unacceptable risk from lighting failure, and emergency lighting should be provided in any area of the premises (not just the escape routes) to reduce that risk.

o If the lighting final circuits do not correspond with the fire compartmentation, then the non-maintained emergency lighting may not operate when required and maintained emergency lighting should be used to reduce that risk.

p The normal level of illuminance for emergency lighting to cover all risks, including use by older people and the presence of obstructions, is a minimum of 1 lux[1] along the centre line of escape routes.

q If there are potential obstructions on the escape route such as stair treads, barriers and walls at right angles, then BS 5266 advises that they should be light in colour against a contrasting background. This contrast may not be appropriate, therefore higher emergency illuminances of greater than 1 lux[1] (0.2 lux) can be installed to reduce the risk due to obstructions.

r If there is likely to be a presence of high physical risk, then a further increase in emergency illuminance and a rapid response time will reduce that risk. The illuminance on the reference plane (note this is not necessarily the floor) should be not less than 10 per cent of the normal illuminance or 15 lux, whichever is the greater. It shall be provided within 0.5 s of failure of the normal lighting supply and continue for as long as the hazard exists (see ICEL 1006 and Chapter 5: Lighting Maintenance, this publication).

s If there is the possibility of arson, then intruder and fire detection and alarm systems in addition to appropriate emergency lighting will reduce that risk.

t If the length of the escape route is excessive, taking into account the fire risk involved and the number of people using the escape route, then the emergency lighting and signage should be assessed. To assess the escape route it may be helpful if the people are timed in escaping from the building during a fire practice in mains failure conditions. If necessary, higher illuminance or repeat signage may reduce the escape time.

u If people such as the public or temporary workers unfamiliar with the layout of the building are likely to be present then higher illuminance or more signs may be required.

v If the escape route passes through open areas, emergency lighting and signage should be installed (see ICEL 1006). A minimum emergency illuminance of 1 lux[1] should be provided along the centre line of the escape route.

w If an area is larger than 60 m², emergency lighting and signage should be installed (see ICEL 1006). A minimum emergency illuminance of 0.5 lux[1] is required for the core area.

x If five or more people are employed in the premises, it is recommended that the emergency lighting should be provided by an installation of fixed luminaires which are automatically switched on upon failure of the normal lighting supply.

y Allowances may need to be made for emergency luminaires that do not operate when called upon. Additional luminaires may be required. Spacing tables are available from manufacturers, and designing in accordance with those spacing tables will ensure the required lighting levels are achieved.

z Emergency lighting should generally be positioned between 2 and 2.5 m above floor level.

Advice on the suitability and location of the escape routes can be obtained from the local fire authority or by consulting the appropriate Home Office guide.

1 0.2 lux is still allowable in the UK as an A deviation in BS 5266 Pt.7:BS EN 1838, for permanently unobstructed escape routes

© The Institution of Engineering and Technology

The recommendations for emergency lighting are detailed in BS 5266: Part 1: 1999. This standard is essential for anyone designing an emergency lighting system, and the guidance given in this Chapter is based upon that given in BS 5266. Further information is given in the ICEL (Industry Committee for Emergency Lighting) documents ICEL 1006 *Emergency lighting design guide* and ICEL 1008 *Emergency lighting – Risk assessment guide* and the ICEL *Checklist for emergency lighting*.

7.5 Siting of escape luminaires

An escape lighting luminaire must be installed at each exit door including emergency exit doors and at any other location that will aid escape, facilitate initiation of alarm, and identification of fire equipment. Emergency lighting luminaires must be installed:

▶ at a final exit door and an emergency exit door to provide illumination of the escape route. Escape route corridors must be illuminated to at least 1 lux on the centre line of the escape route (Figure 7.2a)
▶ within 2 m of a junction of corridors (Figure 7.2b)
▶ within 2 m horizontal distance of a change of direction in an escape route (Figure 7.2c)
▶ within 2 m horizontal distance of a change in floor level or any stair; each tread of the stairs must receive direct light (Figure 7.2d)
▶ externally, within 2 m horizontal distance of any final exit; the purpose is to provide sufficient light for a roll call (Figure 7.2e)
▶ within 2 m horizontal distance of a fire alarm call point, fire fighting equipment and a first-aid point. To illuminate safety signs and equipment (Figure 7.2f)
▶ to illuminate an escalator; an escalator should not be used as an escape route, but nonetheless requires illumination to protect people on it should the normal lighting fail (Figure 7.2g)
▶ in any toilet exceeding 8 m² in area or where natural light is not present; all toilets for disabled people should be provided with emergency lighting (Figure 7.2h)
▶ to provide emergency illumination in any lift (Figure 7.2i)
▶ to provide emergency illumination in a control room such as a motor generator control and plant room and a switchroom (Figure 7.2j)
▶ to illuminate (i) an open area, (ii) an open area with a particular hazard, (iii) an open area with an escape route passing through it and (iv) an open area larger than 60 m²; open areas must be illuminated to at least 1 lux average and 0.5 lux in the central core of the area to within 0.5 m of the walls (Figure 7.2k)
▶ to illuminate an area of high risk; such an area should be illuminated to at least 10 per cent of the normal lighting level or 15 lux, whichever is greater; the risk assessment may identify the need for higher levels of illumination (Figure 7.2l).

Additional emergency lighting should be provided for covered car parks along the normal pedestrian routes.

7.6 Exit signs

Signs should be in accordance with the *European Signs Directive* and be either back illuminated or have an emergency luminaire situated within 2 m horizontal distance. Signs should be the same format throughout the building.

The size of the sign is dictated by the maximum distance at which it will be viewed. The distance is $200 \times H$ for an internally illuminated sign and $100 \times H$ for an externally illuminated sign, where H is the height of the pictogram.

a final exit door

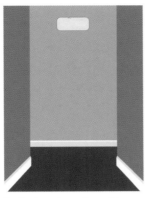

b junction of corridors

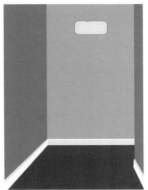

c change of direction

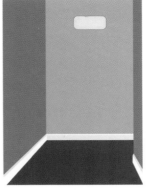

> **Figure 7.2** Siting of emergency luminaires

d change of floor level or stair

e external exit

f fire alarm call point

g escalator

h toilet

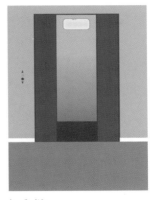

i lift

j control room

k open areas

l area of high risk

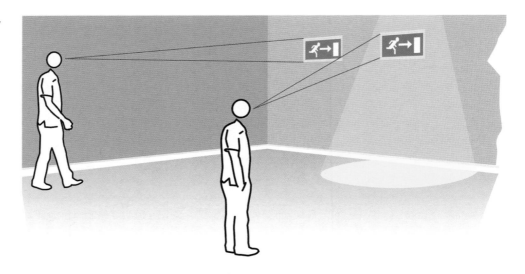

Figure 7.3 Emergency signs should generally be positioned 2–2.5 m above floor level

Figure 7.4 Emergency sign illumination
a from an external luminaire
b internally illuminated

7.6.1 Viewing distance

Emergency signs and emergency lighting should generally be positioned between 2 and 2.5 m above floor level. Signs or luminaires at a higher level may well not be visible should the room start to fill with smoke. All exit, emergency exit and escape route signs need to be illuminated so that they are legible at all times. In the event of failure of the normal electricity supply, such signs should remain illuminated. The illumination should be:

a From an external luminaire and the legend should comply with the Health and Safety Executive recommendations reproduced in Figure 7.4a.

b From an internally illuminated sign, generally constructed in accordance with BS 5499 Part 1. Legends in accordance with BS 5499 will meet the requirements of the new regulations provided they continue to fulfil their purpose effectively. However, the markings should generally be as in Figure 7.4b.

c Self-luminous (e.g. tritium tube) signs in accordance with BS 5499 Part 2, but again with the legend as recommended in the Health and Safety Executive guidance.

7.7 Installation requirements

The installation is required to comply with any statutory requirements applicable to the particular building and with any local regulations. Electrical installations must meet the requirements of BS 7671 *Requirements for Electrical Installations* (as amended).

7.7.1 Luminaires

An emergency lighting luminaire may be either self-contained, that is take its supply from the normal electricity supply when it is available but using internal batteries on the failure of the supply, or supplied from a central battery system. Every luminaire should be constructed in accordance with relevant British standards and be suitable for the environment.

7.7.2 Wiring systems

Cables should be routed through areas of low fire risk. It may be possible to reduce the fire protection of cables where they follow routes of very low fire risk and such areas contain a sprinkler system.

The cable bringing a supply to a self-contained luminaire is not considered to be part of the emergency lighting circuit. This means that such wiring does not have to comply with the requirements for cables given in BS 5266.

Other cables employed in an emergency lighting system must either have an inherently high resistance to attack by fire or must be provided with additional fire protection or must give equivalent protection. Such cables would include those connecting escape lighting (sign or escape route illumination) to a central battery or generating system.

Cables with inherently high resistance to attack by fire are:

a mineral-insulated copper-sheathed cable to BS 6207-1. Such cable may be installed with or without an overall PVC sheath

b cable complying with category B of BS 6387 *Specification for performance requirements for cables required to maintain circuit integrity under fire conditions.*

Wiring systems requiring additional fire protection:

a PVC-insulated cables in accordance with BS 6004 in rigid PVC conduit. Rigid PVC conduit should be of classification 405/100000 or 425/100000 of BS 6099-2-2:1982. (Note that conduit manufactured to BS EN 50086-2-1 will have a different classification number)

b PVC-insulated cables in accordance with BS 6004 installed in steel conduit

c PVC-insulated and sheathed steel wire armoured cable in accordance with BS 6346 or BS 5467.

Additional fire protection may be present if cables are, for example, buried in the structure of the building or situated where there is a negligible fire risk and separated from any significant risk by a wall, partition or floor having at least one hour fire resistance as ascertained by compliance with any of the following:

i specifications tested or assessed under the appropriate Part of BS 476

ii appropriate British Standard specifications or codes of practice

iii specifications referred to by Building Regulations applicable to the building

iv cables enclosed in steel conduit complying with the tests given in BS 6387 for fire resistance.

Where appropriate, requirements for stability, integrity and insulation must be met. The test by fire is applied to the side of the construction remote from the cable. In certain premises a longer duration of fire resistance may be necessary for escape purposes.

7.7.3 System supply
The normal supply should be so arranged that continuity of supply is assured. If, for example, it is the practice to switch off the supply to a premises when not in use or at night, the design of the installation should be such that the supply to the emergency lighting is maintained. This is essential to ensure that emergency lighting batteries remain charged, and are not run down by their supply being disconnected for long periods. This is, of course, as necessary for self-contained luminaires as for a central battery system.

7.7.4 Isolator switches
To reduce the likelihood of the inadvertent disconnection of the supply to the emergency lighting system, isolator switches and protective devices should be installed in a location inaccessible to unauthorised persons, or should be protected against

unauthorised operation. Inadvertent disconnection might result in the rundown of the emergency batteries or their premature failure. Each isolator switch, protective device, key and operating device should be marked:

Emergency   or   **Escape**   or   **Standby lighting**

It is important to give careful consideration to the protective measures selected for emergency lighting supplies. Continuity of supply is particularly important, so precautions must be taken against unwanted operation of a circuit protective device, e.g. at switch-on due to surges or as a result of earth leakage.

7.7.5 Supplies for safety services

Chapter 56 of BS 7671 *Requirements for Electrical Installations* gives specific requirements for supplies for safety services which are summarised in Table 7.2. Refer to the regulations in BS 7671 for full details.

▶ **Table 7.2** BS 7671 requirements for suppliers to safety services

Requirement	Regulation in BS 7671
Supply to be of adequate duration	561-01-01
Fire resistance of adequate duration to be provided	561-01-02
Automatic disconnection preferably not to occur upon first fault	561-01-03
Except for equipment individually supplied by a self-contained battery, the source is to be installed as fixed equipment, suitably placed, ventilated, accessible only to skilled or instructed persons and not used for other purposes	562-01-02 to 561-01-06
Circuit to be separated and have adequate fire resistance	563-01-01, 563-01-02
Protection against overload may be omitted	563-01-03
Switchgear and controlgear identified and accessible only to skilled or instructed persons	563-01-05
Every alarm, indication and control device identified	563-01-06
Circuits must not affect each other	564-01-01
Sources not capable of being operated in parallel	565
Sources capable of being operated in parallel	566

7.7.6 Certification and log book

On completion of an emergency lighting system, or part of a system, or upon completion of a major alteration or addition, a completion certificate should be supplied to the owner or occupier of the premises. A copy of this certificate may be required by the enforcing authority.

A log book should be made available and should be kept on the premises and be readily available for examination by any duly authorised person. The log book should be used to record the following information:

i date of any completion certificate including any certificate relating to alterations
ii date of each periodic inspection and test certificate
iii date and brief details of each service, inspection or test carried out
iv dates and brief details of any defects and of remedial action taken
v date and brief details of any alterations to the emergency lighting installation.

Note: the log book may also include pages relating to other safety records, e.g. fire alarms. The log book may also include details of replacement components for luminaires such as lamp type, battery, fuse rating and type.

7.8 Maintenance

Maintenance of an emergency lighting system should be carried out on a daily, monthly, six-monthly and three-yearly basis. The results of these maintenance activities should be recorded in the log book and kept available for examination. A certificate should be issued for the monthly, six monthly and three-yearly periodic inspections.

7.8.1 Daily activities
i any fault recorded in the log book has been actioned
ii every lamp in a maintained system is lit
iii the main control or indicating panel of a central battery system or engine-driven plant indicates a healthy state
iv any fault is recorded in the log book and the action taken noted.

7.8.2 Monthly
Monthly maintenance is listed below and the results should be recorded on a monthly inspection and test certificate (see Figure 7.5):

i Each self-contained luminaire and internally illuminated exit sign should be energised from its battery by simulation of a supply failure for a sufficient period during which checks should be made to ensure that each lamp is illuminated.
  The period of the simulated supply failure should be as short as possible to avoid discharging the battery and should under no circumstances exceed one quarter of the rated duration of the luminaire or sign. At the end of the test the supply to the normal lighting should be restored and a check made that any indicator lamps indicate a healthy state.
ii A central battery system should be energised from its central battery by simulation of a supply failure for a sufficient period during which checks should be made to ensure that each lamp is illuminated. The period of the failure should be minimised and should not exceed a quarter of the rated duration of the battery. If it is not possible to visually examine all the luminaires and signs during this period, further tests will have to be made after the batteries have been fully recharged.
  At the end of the test the supply should be returned to normal and charging arrangements checked.
iii Engine-driven generating plants should be started by simulation of mains failure and allowed to energise emergency lighting for a period of at least one hour.
iv For engine-driven generating plant with back-up batteries, the system should be tested to check the battery supply, by simulation of failure of the supply with the

Emergency lighting system monthly service		
Owner/occupier		
Address of premises		**Postcode**
System description		
Date of inspection and test		
System verifier's name		
System verifier's address		**Postcode**
Signature		**Date**
1 Defects recorded in the log book have been corrected	☐ yes	☐ no
2 Luminaires and signs are clean	☐ yes	☐ no
3 Luminaires and signs operate correctly upon simulated supply failure	☐ yes	☐ no
Engine-driven generator		
4 Engine-driven generator operates correctly	☐ yes	☐ no
5 Manufacturer's recommended maintenance has been carried out on engine driven generator	☐ yes	☐ no
6 Fuel tanks, oil and coolant levels have been checked and topped up as necessary	☐ yes	☐ no
Battery systems		
7 Level of electrolyte in batteries of central battery systems and generator starter batteries have been checked	☐ yes	☐ no
Final checks		
8 All indicator lamps are functioning	☐ yes	☐ no
9 Data has been recorded in log book	☐ yes	☐ no

▶ **Figure 7.5** Emergency lighting system monthly service

engine-driven generating plant prevented from starting. After this check, the starting of the engine should be allowed in the normal manner, and the emergency lighting system run from the generating plant for at least one hour. At the end of the test period the emergency equipment should be returned to normal, and healthy indications checked. Fuel tanks should be filled, oil and coolant levels topped and other maintenance activities associated with the prime mover carried out as necessary.

7.8.3 Six-monthly

At six-monthly intervals (see Figure 7.6), the monthly test should be carried out plus:

i Each three-hour self-contained luminaire and internal illuminated sign should be energised from its battery for one hour by simulation of a supply failure. Each one-hour self-contained luminaire and internal illuminated sign should be energised from its battery for 15 minutes by simulation of a supply failure. During the period of the simulated failure, all luminaires and/or signs should be checked to ensure that each lamp is illuminated.

At the end of the test the supply should be returned to normal and charging arrangements checked.

ii Each three-hour central battery system should be continuously energised from its battery for one hour by simulation of a supply failure. Each one-hour central battery system should be continuously energised from its battery for 15 minutes by simulation of a supply failure. During the period of the simulated failure, all luminaires and/or signs should be checked to ensure that each lamp is illuminated.

At the end of the test the supply should be returned to normal and charging arrangements checked.

iii Engine-driven plant should be tested in accordance with the monthly schedule.

iv The engine of each engine-driven generating plant with back-up battery should be prevented from starting. The emergency lighting system should then be energised solely from the back-up battery for a continuous period of one hour by simulation of supply failure.

The starting system of the engine should then be restored to normal operation and the engine allowed to start up in the normal way to energise the emergency lighting system for a further continuous period of one hour.

The system should be returned to normal. Fuel tanks should be filled, oil and coolant levels topped up as necessary.

7.8.4 Three-yearly

The six-monthly inspection should be carried out with the following additional tests:

i The emergency lighting installation should be tested and inspected to ascertain compliance with the recommendations of BS 5266 and a periodic inspection and test certificate issued.

ii Each self-contained luminaire or internally illuminated sign should be tested for its full duration.

At the end of the test period the supply to the normal lighting should be restored and any indicator lamp or device checked to ensure that it is showing that the normal supply has been restored.

iii Each central battery system should be tested for its full duration.

At the end of the test period the supply to the normal lighting should be restored and any indicator lamp or device checked to ensure that it is showing that the normal supply has been restored. The charging arrangements should be checked for proper functioning.

Emergency lighting system six-monthly service		
Owner/occupier		
Address of premises		
		Postcode
System description		
Date of inspection and test		
System verifier's address		
		Postcode

Signature	**Date**	

1	Defects recorded in the log book have been corrected	☐ yes	☐ no
2	Luminaires and signs are clean	☐ yes	☐ no
3	Luminaires and signs operate correctly upon simulated supply failure	☐ yes	☐ no
4	Each 3 h self-contained luminaire and internally illuminated sign has been energised from its battery for 1 h and operates correctly	☐ yes	☐ no
5	Each 1 h self-contained luminaire and internally illuminated sign has been energised from its battery for 15 min and operates correctly	☐ yes	☐ no
6	Supply has been properly restored following tests in 3, 4 and 5 above	☐ yes	☐ no
7	Charging arrangements functioning	☐ yes	☐ no

Engine-driven generators

8	Engine of engine-driven plant has been prevented from starting	☐ yes	☐ no
9	Emergency lighting system energised solely from back-up battery for 1 h by simulation of supply failure	☐ yes	☐ no
10	Luminaires and signs operate correctly	☐ yes	☐ no
11	Starting system restored and engine allowed to start up and energise emergency lighting system for a period of 1 h.	☐ yes	☐ no
12	Luminaires and signs operate correctly	☐ yes	☐ no
13	System restored to normal	☐ yes	☐ no
14	Manufacturer's recommended maintenance has been carried out on engine driven generator(s)	☐ yes	☐ no
15	Fuel tanks, oil and coolant levels have been checked and topped up as necessary	☐ yes	☐ no

Battery systems

16	Level of electrolyte in batteries of central battery systems and generator starter batteries have been checked	☐ yes	☐ no

Final checks

17	All indicator lamps are functioning	☐ yes	☐ no
18	Data has been recorded in log book	☐ yes	☐ no

Electrical Maintenance
© The Institution of Engineering and Technology

iv Each generator back-up battery should be tested for its full duration.

At the end of the test period the system should be restored to normal operation and the charging arrangements for the back-up and engine starting batteries checked for proper functioning. Any indicator lamp or device should be checked to ensure that it is showing that normal arrangements have been restored. Fuel tanks, oil coolant levels topped up etc. as necessary.

7.8.5 Subsequent annual test

For self-contained luminaires with sealed batteries, after the first three-yearly test, the three-yearly test should be carried out annually.

◀ **Figure 7.6** Emergency lighting system six-monthly service

Emergency lighting periodic inspection and test certificate
Owner/occupier
Address of premises
Postcode
System description
Date of inspection and test
System verifier's address
Postcode

I/we hereby certify that the emergency lighting installation at the above premises has been inspected and tested in accordance with the schedule below by me/us and to the best of my/our knowledge and belief complies at the time of my/our test with the recommendations of BS 5266 *Emergency lighting* Part 1: 1999 *Code of practice for the emergency lighting of premises other than cinemas and certain other specified premises used for entertainment*, published by BSI, for a category ....* installation, except as stated below.

* Enter M/1, 2 or 3 or NM/1, 2 or 3 as appropriate

Signature	**Date**

Qualification†

Variations
Details of variations from the code of practice (BS 5266: Part 1: 1999)

† Qualifications: a suitably qualified electrical engineer or a member of the Electrical Contractors' Association or the Electrical Contractors' Association of Scotland; or a certificate holder of the National Inspection Council for Electrical Installation Contracting; or a qualified person acting on behalf of one of these (in which case it should be stated on whose behalf he is acting). Where acceptable to the enforcing authority the authorised representative of a manufacturer of emergency lighting equipment may be deemed to be a suitably qualified person.

▸ **Figure 7.7** Emergency lighting periodic inspection and test certificate

Page 1 of 3

Results of inspection and tests		
1 Are the entries made in the log book correct?	☐ yes	☐ no
2 Are record drawings available?	☐ yes	☐ no
3 Are record drawings correct?	☐ yes	☐ no
4 Signs		
Are the signs correctly positioned? (6.9)	☐ yes	☐ no
Are details of the signs correct? (6.9)	☐ yes	☐ no
Do the self-luminous signs (if any) need changing before the date of the next scheduled inspection? If so state date .... (see label on sign) (6.9)	☐ yes	☐ no
5 Luminaires		
Are luminaires correctly positioned? (6.7, 6.8 and 10.3)	☐ yes	☐ no
6 Illumination for safe movement (clause 5 and see record drawings)		
Are the correct lamps installed in the luminaires? (6.13)	☐ yes	☐ no
Has there been any change in the decor or layout of the premises since the last inspection, which has caused any significant reduction in the effectiveness of the lighting system?	☐ yes	☐ no
(Any changes to be stated under comment below.) Is the installation in a generally satisfactory condition?	☐ yes	☐ no
7 Marking		
Are the category and nominal operating voltage of the system clearly marked or readily identifiable? (6.13)	☐ yes	☐ no
Are luminaires clearly marked to indicate the correct lamp for use? (6.13)	☐ yes	☐ no
Is information available to ensure correct battery replacement? (6.13)	☐ yes	☐ no
8 Wiring systems (clause 8)		
Are the results recorded on the last inspection and test certificate satisfactory?	☐ yes	☐ no
State the date of this inspection and test	....	

9 Power services		
Are the charging arrangements for batteries satisfactory? (6.11, 12.2 and 12.4)	☐ yes	☐ no
Do changeover devices operate satisfactorily upon simulation of failure of the normal supply? (6.11 and 12.4)	☐ yes	☐ no
10 Central battery systems including backup batteries		
After operation for the rated duration:		
i Do all luminaires operate? (6.7, 6.8 and 12.4);	☐ yes	☐ no
ii Are all signs illuminated and visible? (6.9 and 12.4);	☐ yes	☐ no
Following the restoration of the system to normal:		
i Is the battery charger functioning? (6.11 and 12.4);	☐ yes	☐ no
ii Are the levels and the specific gravities of the battery electrolytes satisfactory, where applicable?	☐ yes	☐ no
11 Engine-driven generating plant		
a After a period of operation of at least 1 h:		
i Do all luminaires operate? (6.7, 6.8 and 12.4);	☐ yes	☐ no
ii Are all signs illuminated and visible? (6.11 and 12.4);	☐ yes	☐ no
iii Does the backup battery, where installed, operate satisfactorily? (see above)	☐ yes	☐ no
b Following the restoration of the system to normal:		
i Is the battery charger for the engine starter functioning? (6.11 and 12.4)	☐ yes	☐ no
ii Are the fuel, coolant and lubricating oil levels correct? (12.4)	☐ yes	☐ no
12 Self-contained luminaires and signs		
After operation for the rated duration, does each self-contained luminaire and sign operate? (6.9, 6.11 and 12.4)	☐ yes	☐ no
Following restoration of the system to normal supply, is the battery charger functioning? (6.11 and 12.4)	☐ yes	☐ no
13 Comment (if any) and variations (if any) from the code of practice.		

1 Because of the possibility of failure of the supply to the normal lighting occurring shortly after a period of testing, all tests should be undertaken at times of minimum risk. Alternatively, suitable temporary arrangements should be made until the batteries have been recharged.

2 The figures in brackets indicate the relevant clauses in BS 5266: 1999.

Page 3 of 3

Fire detection and alarm systems

8.1 British Standards

The applicable British standards for fire detection and alarm systems in buildings and dwellings are given in Table 8.1.

Persons with responsibilities for design, installation or maintenance of fire detection and alarm systems in buildings and dwellings should obtain a copy of the appropriate standard.

Buildings	BS 5839-1: 2002	Fire detection and fire alarm systems for buildings – Part 1: Code of practice for system design, installation, commissioning and maintenance
Dwellings	BS 5839-6: 2004	Fire detection and fire alarm systems for buildings – Part 6: Code of practice for the design, installation, and maintenance of fire detection and fire alarm systems in dwellings

▶ **Table 8.1** British Standards for fire detection and alarm systems in buildings

8.2 Buildings

8.2.1 Fire detection and alarm systems for buildings
In England, Wales and Northern Ireland, requirements for effective means of giving warning in case of fire are incorporated within the *Building Regulations*. Certain premises in Scotland are required to have fire alarm systems under the Regulations in Scotland.

The *Building Regulations* apply both to new buildings and to extensions or material alterations to existing buildings. The requirement from Part B of Schedule 1 to the *Building Regulations 2000* is given in Table 8.2.

Requirement	Limits on application
Means of warning and escape	
B1. The building shall be designed and constructed so that there are appropriate provisions for the early warning of fire, and appropriate means of escape in case of fire from the building to a place of safety outside the building capable of being safely and effectively used at all material times.	Requirement B1 does not apply to any prison provided under section 33 of the *Prisons Act* 1952 (power to provide prisons etc.).

▶ **Table 8.2** Requirement from Part B, Schedule 1 to the Building Regulations

Approved Document B states 'In the Secretary of State's view the requirement of B1 will be met if:

a there are routes of sufficient number and capacity, which are suitably located to enable persons to escape to a place of safety in the event of fire
b the routes are sufficiently protected from the effects of fire by enclosure where necessary
c the routes are adequately lit
d the exits are suitably signed, and if
e there are appropriate facilities to either limit the ingress of smoke to the escape route(s) or to restrict the fire and remove smoke; all to an extent necessary that is dependent on the use of the building, its size and height
f there is sufficient means for giving early warning of fire for persons in the building.'

In England, Scotland and Wales, the *Fire Precautions (Workplace) Regulations 1997* (as amended) require that where necessary to safeguard employees in the case of fire, workplaces must be equipped with appropriate fire detection and alarm systems. Premises that require a fire certificate under the *Fire Precautions Act 1971* must also be provided with adequate means of giving warning in case of fire. Equivalent legislation applies in Northern Ireland.

8.2.2 Need for a fire alarm system

The authority responsible for enforcing fire safety or the result of a risk assessment carried out by the owner, landlord, occupier or employer will determine whether a fire alarm system is required and the category of that system. Except in very small premises, most buildings will require some form of fire alarm system. The insurance company may require that a suitable automatic fire alarm system be fitted and adequately maintained. A manual system can be sufficient for premises where persons do not sleep. Where people do sleep, an automatic system with additional manual call points will probably be required.

Information on fire precautions is given in the various parts of BS 5588: *Fire precautions in the design, construction and use of buildings*.

8.2.3 Categories of system

Fire alarm systems are installed in buildings to satisfy one or both of:

i Protection of life
ii Protection of property (which may also include protection against business interruption and protection of the environment).

Fire alarm systems are divided into the categories given in Table 8.3.

A system that is intended to fulfil the objectives of more than one Category of system needs to comply with the recommendations for each Category. Where the objectives of more than one type of system are to be satisfied, the system should be described as a Category X/Y system (e.g. L2/P2).

The Category of system needs to be defined in the specification and, except for Category L1 or P1 systems, the details of the areas of the building to be protected. Table 8.4 gives the Category of system that is typically installed in various types of premises.

Category		Description and objectives
Manual fire detection systems		
M		manual system; no automatic fire detectors
Automatic fire detection systems		
L The objective of Category L systems is the protection of life	L1	systems installed throughout all areas of the building with the objective of giving the earliest possible warning of fire allowing the longest possible time for escape
	L2	systems installed only in defined parts of the building with the objectives of (i) giving warning of fire to enable all occupants to escape safely before the escape routes are impassable and (ii) affording early warning of fire in specified areas of high fire hazard and/or high fire risk
	L3	systems installed only in defined parts of the building with the objective of giving warning of fire to enable all occupants to escape safely before the escape routes are impassable
	L4	systems installed in escape routes including circulation areas and circulation spaces such as corridors and stairways with the objective of enhancing the safety of occupants by providing warning of smoke within escape routes
	L5	systems in which the protected areas and/or the locations of detectors are designed to satisfy a specific fire safety objective other than that of a Category L1, L2, L3 or L4 system
P The objective of Category P systems is the protection of property	P1	systems installed throughout all areas of the building with the objective of giving the earliest possible warning of fire so as to minimise the time between ignition and the arrival of fire fighters
	P2	systems installed only in defined parts of the building with the objective of providing early warning of fire in areas of high fire hazard or areas in which the risk to property or business continuity from fire is high

▶ **Table 8.3** Categories of fire alarm systems

Table 8.4 Choice of appropriate Category of fire detection and alarm system. This table describes the category of system that is typically installed in various types of premises. The information in the table is not intended to constitute recommendations, but simply provides information on custom and practice

Types of premises	Typical category of system	Comments
Offices Shops Factories Warehouses Restaurants	M, P2/M or P1/M	a Category M system normally satisfies the requirements of legislation. It is often combined with a Category P system to satisfy the requirements of insurers or as company policy or to protect against business interruption
Hotels Hostels	L1 or L2	bedroom areas usually fitted with a Category L3 system; detectors are normally installed in all other areas, as a fire could pose a threat to sleeping occupants, making the system at least L2
Large public houses without residential accommodation	M	
Public houses with residential accommodation	L2	
Schools (other than small single storey schools with less than 160 pupils)	M or M/P2 M/P2/L4 M/P2/L5	
Hospitals	L1 (with possible minor variations)	detailed guidance on areas to be protected and possible variations is given in HTM 82

Places of assembly:
  cinemas
  theatres
  nightclubs
  exhibition halls
  museums
  galleries
  leisure centres
  casinos
	M (for small premises e.g. accommodating less than 300 persons),
	
	L1 to L4 (for other premises)
		L1 systems are often provided in large or complex buildings

Transportation terminals	M/L5	
Covered shopping centres	L1 to L3	system design normally needs to be part of a fire engineering solution
Residential care homes	L1 to L3	L1 is appropriate for large care homes
Prisons	M/L5	
Phased evacuation buildings	L3	
Buildings in which other fire precautions such as means of escape depart from recognised guidance	M/L5	automatic fire detectors should be sited so as to compensate for the lower standard in other fire precautions

▸ **Table 8.4** *continued*

Types of premises	Typical category of system	Comments
Buildings with inner rooms from which escape is only possible by passing through another (access) room	M/L5	smoke detectors should be sited in the access room
Buildings in which the automatic fire detection system is required to operate other fire protection systems such as magnetic door holders	M/L5	automatic fire detectors must be suitably located to ensure doors are closed when smoke is present
Situations in which fire could readily spread from an unoccupied area and prejudice means of escape from occupied areas	M/L4 or M/L5	plant rooms and storage areas need not necessarily be fitted with automatic detectors
Buildings in which automatic fire detection is provided as a requirement of a property insurer	M/P1 M/P2	

▸ **Table 8.5** System components

System components	
Manual call points	BS EN 54-11 for Type A (single action) manual call points
Point heat detectors	BS EN 54-5 for Class A1 or A2 detectors (ambient temperature below 40 °C)
	BS EN 54-5 for Class B–G detectors (ambient temperature above 40 °C)
Point smoke detectors	BS EN 54-7
Flame detectors	BS EN 54-10
Optical smoke beam detectors	BS EN 54-12
Carbon monoxide fire detectors	in the absence of any international, European or British standard should meet the requirements of LPS 1265
Control and indicating equipment	BS EN 54-2
Audible fire alarm devices	BS EN 54-3
Power supply equipment	BS EN 54-4
Cables	see the section on cables in this chapter
Software controlled control and indicating equipment	BS EN 54-2

8.2.4 System specification

The user or purchaser of the fire alarm system, or their consultant, should ensure that, as necessary, the authority responsible for enforcing fire safety is consulted (e.g. building control authority, fire authority, local authority, HSE). The property insurer should also be consulted.

8.2.5 Design of system

The fundamental design consideration for a fire detection and alarm system is whether life or property or both are to be protected. The System Category must next be defined. Detailed information on design of fire detection and alarm systems is given in section 2 of BS 5839-1: 2002.

8.2.6 System components

System components should correspond to the standards given in Table 8.5.

8.2.7 Fault monitoring

A fault on a fire detection and alarm system could prevent the system from giving an alarm. In addition, work may need to be performed on the system during which it may not be fully operating. Faults will occur occasionally and are most likely to occur in the equipment and wiring external to the control and indicating equipment.

Critical signal paths should be both protected against fire and mechanical damage and monitored. Compliance with the recommendations for maintenance will ensure that any fault is quickly repaired such that the probability of a fault existing at the time of a fire is extremely low. In the event of a fire, providing the recommendations of BS 5839-1 have been met, fire damage to the wiring during the evacuation period is unlikely to occur.

In addition, if the system is used to actuate other fire protection systems or safety facilities, reference should be made to BS 7273 or other applicable codes of practice. Where tactile fire alarm devices are provided for people with impaired hearing, the system should indicate any failure to receive a monitoring signal correctly.

8.2.8 System integrity

Fire alarm systems should be designed to meet the recommendations of Table 8.7.

8.2.9 False alarms

False alarms cause disruption to work, create a drain on fire service resources and may prevent people reacting correctly when there is a real fire ('Oh – it's just another test or a false alarm'). See Table 8.8 which lists types of false alarm.

The numbers of false alarms are increasing. All parties with responsibilities for fire alarm systems must take action to limit the number of false alarms at all stages including design, installation, commission, verification, use and maintenance.

False alarms only occasionally result from equipment problems. Most false alarms are initiated by fire-like phenomena in the building (such as burnt toast, cooking, steam, or people smoking), or inappropriate action by people in the building or accidental damage. False alarms also arise due to malicious intent, on the one hand, and due to 'good intent' on the other where a person genuinely believes there is a fire.

Maximum time period within which fault must be indicated	Condition under which fault is required to be indicated	
Within 100 seconds	short-circuit or open-circuit in any circuit connecting:	manual call point
		fire detector
		fire alarm sounder
		a power supply in a separate enclosure and the equipment it is powering
		separated control and indicating equipment
		separated main and repeat control and indicating equipment such as a mimic diagram
		separated equipment for transmission of alarm signals to an alarm receiving centre
		the fire detection system and the voice alarm system
	removal of one or more of the following detachable items	manual call point
		fire detector
		fire alarm sounder
	any earth fault that would prevent proper system operation	
	the operation of any protective device that would prevent proper system operation	
Within 15 minutes	failure of the standby power supply	
	disconnection of any one battery where the supply comprises a number of batteries connected in parallel	
Within 30 minutes	failure of the main power supply (a.c. input) to any part of the system	
	failure of the battery charger	
	reduction of the battery voltage to less than the voltage specified in BS EN 54-4 at which a fault warning must be given	

▶ **Table 8.6** Fault monitoring

Table 8.7 System integrity

System integrity		
Circuits	a fault on one circuit connecting call points, detectors and sounders should not affect any other circuit	
	a fault between a detector and sounder circuit should not affect any other circuit	
	two simultaneous faults on a call point or detector circuit should not disable protection over an area greater than 10,000 m²	
	a single short-circuit or open-circuit on a detector circuit:	should not disable protection over an area of more than 2000 m² and
		should not disable protection on more than one floor of the building plus five devices on the floor above and five on the floor below
	a short-circuit or open-circuit on a sounder circuit:	should not stop at least one sounder sited close to the control and indicating equipment operating; this sounder should have an identical sound to the other sounders
	duplicated power circuits should be provided between power supply equipment and control and indication equipment which is physically separated such that a single short-circuit or open-circuit does not completely remove power from the control and indicating equipment	
Detachable detectors	removal of a detector should not affect the operation of any call point. The possibility of malicious removal should be considered	
Removable detectors or call points	removal of a call point or detector from its circuit should not affect the ability of any sounder to respond to an alarm signal except for devices incorporating a sounder and detector in the same unit	
Sounders	sounders should only be able to be removed with a special tool	
Disablement of part of the system	where disablement of part of the system is possible (for maintenance purposes)	it should be possible to disable protection in one zone without affecting any other zones
		a control to permit evacuation of the building should be provided close to the control and indicating equipment
Software controlled systems	for such systems with more than 512 detectors and/or call points, reference should be made to the manufacturer's instructions as to how to achieve compliance with BS EN 54-2	
Buildings designed to accommodate the public in large numbers	at least two sounder circuits should be provided in every uncompartmented public space if: ▸ the space is more than 4000 m² or ▸ the space is designed to accommodate more than 500 members of the public	

Types of false alarm	Typical causes
False alarm due to fire-like phenomena	burnt toast, cooking fumes, steam
False alarm with good intent	where a person genuinely believes there is a fire
Malicious false alarm	deliberate false alarm
False alarm due to equipment problems	equipment failure resulting in a false alarm

▶ **Table 8.8** Types of false alarm

False alarms must be logged including the date and time, the identity and location of the device, the category of false alarm (unwanted, malicious, equipment, good intent) and the reason, the action taken and the name of the person responsible for recording the false alarm. False alarm records should be kept separately, within the log book, from records of other 'events', such as routine testing, servicing and faults.

The number of false alarms per year should be expressed as the number of false alarms per 100 detectors per annum and recorded in the log book. An investigation should be carried out if:

▶ the rate of false alarms over the previous 12 months has exceeded one false alarm per 25 detectors per annum
▶ more than ten false alarms have occurred since the time of the previous service visit
▶ there have been two or more false alarms, other than false alarms with good intent, from any single manual call point or detector or detector location, or
▶ any persistent cause of false alarms can be identified.

The investigation should identify any corrective actions that should then be taken.

The responsible person should ensure that all staff, contractors and others visiting the premises are aware of the alarm system and the precautions necessary to avoid false alarms.

8.2.10 Power supplies
Source of supply
The mains supply for a fire detection and alarm system should be derived via a double pole isolating protective device from the load side of the main isolating device for the building. The fire alarm supply should be dedicated exclusively to the fire alarm system and should not serve any other systems or equipment.

Labelled
Every isolator or protective device that can disconnect the supply to the fire alarm system, other than the main isolator for the building, should be labelled:

FIRE ALARM

in the case of a protective device that serves only the fire alarm circuit but incorporates no switch, or

FIRE ALARM, DO NOT SWITCH OFF

in the case of a switch, whether incorporating a protective device or not, that serves only the fire alarm circuit, or

> WARNING. THIS SWITCH ALSO CONTROLS THE
> SUPPLY TO THE FIRE ALARM SYSTEM

in the case of any switch that disconnects the mains supply both to the fire alarm and to other circuits.

Unauthorised operation

Every isolator, switch or protective device that is capable of disconnecting the mains supply to the fire alarm system should be situated in a position inaccessible to unauthorised persons or be protected against unauthorised operation by means such as requiring a special tool.

RCD protection

The circuit supplying the fire alarm system should not be protected by a residual current device unless this is necessary to comply with the requirements of BS 7671. Where an RCD is necessary for electrical safety, a fault on any other circuit or equipment in the building should not be capable of resulting in isolation of the supply to the fire alarm system.

Power supply unit

Irrespective of the condition of any standby battery, the mains supply should be capable of supplying the maximum load of the alarm system. Transition from normal to standby and vice versa should not cause any interruption to system operation or result in a false alarm. A fault in the normal supply should not adversely affect the standby supply or vice versa. Operation of a single protective device should not result in the failure of both the normal and the standby supplies. The normal and standby supplies should each be independently capable of supplying the maximum alarm load of the system.

Standby supply

The standby supply should consist of a rechargeable battery with an automatic charger. The battery should have a life of at least four years, and automotive batteries – the type used for starting car engines – should not be used. Batteries should be labelled. Further information on battery requirements is given in Clause 25.4 of BS 5839-1.

8.2.11 Cables

Cables used for:

▶ all parts of the critical signal paths (components and connections between every call point and detector and the control panel and between the control panel and the sounder(s))
▶ the final circuit providing low voltage mains supply to the system
▶ the extra low voltage (ELV) supply from an external power supply unit

should be either standard fire resisting cables meeting the PH 30 classification when tested in accordance with EN 50200 or fire resisting cables meeting the PH 120 classification when tested in accordance with EN 50200 and subject to the tests detailed in clauses 26.2 *d* and *e* of BS 5839, respectively. Such cables will be one of the following:

▶ mineral-insulated copper-sheathed cables (MICC) with an overall sheath conforming to BS EN 60702-1 with terminations corresponding to BS EN 60702-2
▶ cables conforming to BS 7629
▶ cables conforming to BS 7846 (steel-wire armoured SWA cables)
▶ cables rated at 300/500 V or greater providing the same degree of safety as that afforded by compliance with BS 7629.

BS 8434-1 2003: *Methods of test for assessment of the fire integrity of electric cables. Test for unprotected small cables for use in emergency circuits* gives further information on cables for fire alarm systems.

Cable supports

Methods of cable support should be such that circuit integrity will not be reduced below that afforded by the cable used and should withstand a similar temperature and duration to that of the cable while maintaining adequate support.

Joints

Circuit integrity should be maintained at any joint or connection. Plastic connectors should be avoided. Unnecessary joints or connections should be avoided wherever possible. Joints should be enclosed within junction boxes labelled:

> FIRE ALARM

Mechanical damage

MICC cables conforming to BS EN 60702 and SWA cables conforming to BS 7846 may be used throughout all parts of the system without additional mechanical protection. Other cables should be protected where mechanical damage or rodent attack is likely. Mechanical protection may be needed in areas less than 2 m above floor level. Fire alarm cables should not be installed within the same conduit as cables of other services. Where fire alarm cables share a common trunking, the fire alarm cable should be installed in a separate compartment in the trunking which should be separated from other compartments by a strong, rigid and continuous partition. Conduit should conform to BS EN 50086 and trunking should conform to BS 4678-4.

Cross-sectional area

All conductors should have a CSA of at least 1 mm^2.

Segregation

Low and extra-low voltage cables in fire alarm systems should be segregated. The mains supply cable to the panel should not enter through the same cable entry as extra-low voltage cables.

Multicore cables

None of the conductors should be used for circuits other than those of the fire alarm system.

Colour

All fire alarm cable should be of a single common colour to enable easy distinction from other cables. The preferred colour is red.

8.2.12 User responsibilities

The owner or other person responsible for the premises should appoint a single named responsible person to supervise the system. The responsible person's duties include ensuring:

▸ the system is maintained in accordance with BS 5839-1
▸ records and operating instructions are kept and made available to any person modifying, repairing or servicing the system
▸ occupants of the premises are aware of their roles and responsibilities.

In addition, the responsible person should ensure:

- the control and indicating equipment is checked at least once every 24 hours to confirm there are no faults on the system
- testing and maintenance is carried out as recommended by BS 5839-6.
- the log book is kept up to date and available
- occupants are instructed in the use of the system and are aware of how to avoid false alarms
- escape procedures are agreed with the fire authority and implemented, including practice procedures, so that people can be safely evacuated in the event of an alarm. All staff are instructed and practised in proper actions to be taken in the event of a fire and staff with particular responsibilities are also suitably trained
- all passageways, paths or other avenues of escape, and access to fire alarm and fire fighting equipment, are kept clear at all times
- actions are taken, as necessary, to limit false alarms
- a clear space of 500 mm is preserved below each detector in all directions.
- manual call points remain unobstructed and conspicuous
- building work, such as structural work and redecoration, does not reduce the level of protection provided by the system
- drawings are updated when changes are made to the system.

8.2.13 Spare parts
The responsible person should ensure the following spare parts are held on the premises:

- six frangible elements and appropriate tools for manual call points unless there are less than twelve manual call points in which case only two spare frangible elements and appropriate tools need be held
- other spare parts as agreed between the user and the organisation responsible for maintaining and/or servicing the system.

8.2.14 Log book
The responsible person should ensure that a log book (see Figure 8.1a) is kept which includes the following information:

a name of the responsible person
b brief details of maintenance arrangements
c dates and times of all alarm signals (false, test, practice and genuine), including the location and type of the device initiating the alarm
d causes, circumstances and category of all false alarms
e dates, times and types of all tests
f dates, times and types of all faults and defects and when rectified
g date and types of all maintenance (e.g. service visit or non-routine attention).

8.2.15 Maintenance
It is important that testing and maintenance of fire alarm systems does not result in false alarms.

Maintenance of fire detection and alarm systems will fall into one of the following categories:

- weekly, monthly and quarterly testing
- periodic inspection and servicing
- work to be carried out each year
- appointment of a new servicing organisation
- modification.

Log book for a fire alarm system
Address of premises ....
Name and address of fire alarm maintenance, repair and modification organisation ....
Responsible person ....
System designed by ....
System installed by ....
System accepted by ....
System verified by ....
The system is maintained under contract by .... until...
Person to be contacted if maintenance is required   Phone number ....
Normal maximum attendance time for a maintenance technician is ....
Expendable component replacement periods (list) ....

▶ **Figure 8.1a** Fire alarm log book, page 1

Alarms and routine maintenance

(Fire alarm signals, weekly tests, monthly and quarterly tests, non-routine attention, faults.) Record all events other than false alarms and maintenance work.

Date and time	Event	Zone	Device	Action required	Date completed	Initials

False alarms

Maintenance work

▶ **Figure 8.1b** Fire alarm log book, page 2

Weekly, monthly and quarterly testing (Figure 8.2)

Most systems will only require a weekly test which will normally be made by the designated responsible person. Monthly and quarterly testing will be necessary for larger systems with vented batteries and a standby generator.

Routine testing permits occupants of the building to become familiar with the alarm signals.

Modern fire alarm systems generally incorporate monitoring, and faults are indicated automatically. A responsible person must routinely check the system, identify any fault indications and ensure they are attended to.

Periodic inspection and servicing (Figure 8.3)

Periodic inspection and servicing should be by an organisation certificated by a United Kingdom Accreditation Service (UKAS)-accredited body with knowledge of fire detection and alarm systems including:

▶ knowledge of the causes of false alarms
▶ sufficient information regarding the system
▶ adequate access to spares.

The system records should always be made available to any person servicing, modifying or repairing the system.

Work to be carried out each year

The form in Figure 8.4 details ongoing work to be carried out on a fire alarm system.

8.2.16 Appointment of a new servicing organisation

Upon appointment of a new servicing organisation, the form below should be filled in and kept with the system records.

8.2.17 Modifications

Every modification made to a fire alarm system should be recorded and the form kept with the system records. The system records should always be made available to any person modifying, repairing or servicing the system.

Weekly, monthly and quarterly test and maintenance of a fire alarm system

▶ The weekly test should be carried out at approximately the same time each week.
▶ The duration of the alarm signal should not exceed 1 min.
▶ Where some employees only work during hours other than when the system is tested, an additional test should be carried out monthly at an appropriate time to ensure these employees are familiar with the system.

Address of premises	
....	

Ancillary outputs isolated (if necessary), such as automatic dialling of call centre	☐ yes ☐ no

Weekly checks

A manual call point should be operated during normal working hours (a different call point should be used each week)	Call point used ....
Control equipment processes fire alarm signal and operates the sounders	☐ yes ☐ no
Alarm receiving centre contacted and advised test is imminent (if appropriate)	☐ yes ☐ no
Alarm receiving centre receives signal	☐ yes ☐ no
Building occupants to report any instance of poor audibility of the alarm system. Comments (if any)	☐ yes ☐ no
Voice alarm system tested (if appropriate)	☐ yes ☐ no
Any faults on system identified and appropriate action taken	☐ yes ☐ no
Entry made in log book	☐ yes ☐ no

Monthly checks

Automatically started generator is started by simulated mains failure and operated on load for one hour Fuel tanks refilled and oil and coolant levels checked and topped up as necessary	☐ yes ☐ no
If vented batteries are used as a standby power supply, the batteries and their connections should be made to ensure they are in good condition and the electrolyte levels are correct	☐ yes ☐ no

Quarterly check

All vented batteries and their connections should be examined by a person competent in battery installation and maintenance technology. Electrolyte levels should be checked and topped up as necessary	☐ yes ☐ no
Test performed by Name .... Sign .... Date	Date of next test ....

▶ **Figure 8.2** Weekly, monthly and quarterly fire alarm testing

Periodic inspection and servicing of fire alarm system

Period between inspections and tests to be determined by risk assessment
Period between inspections and tests not to exceed six months

Address of premises

....

Name and address of servicing organisation

....

....

Test performed by Name Sign	Date of next test ....
Category of system	....

Log book	Log book up to date	☐ yes ☐ no
	Faults recorded in log book have been actioned	☐ yes ☐ no
Visual inspection	Siting of call points, detectors and sounders affected by any structural or occupancy changes	☐ yes ☐ no
	All manual call points unobstructed and conspicuous	☐ yes ☐ no
	Any new exits have an adjacent manual call point	☐ yes ☐ no
	All new or relocated partitions are NOT within 500 mm of any detector	☐ yes ☐ no
	All storage at least 300 mm below any ceiling	☐ yes ☐ no
	Clear space of 500 mm below each detector	☐ yes ☐ no
	Detectors still suitable for detection (no change to the use or occupancy of an area has affected detector operation or has created the potential for false alarms)	☐ yes ☐ no
	New detectors and alarms have been installed in building alterations and extensions	☐ yes ☐ no
False alarms	Records of false alarms checked	☐ yes ☐ no
	Rate of false alarms during the previous 12 months	....
	Action taken in respect of false alarms Action taken	☐ yes ☐ no

▶ **Figure 8.3** Periodic inspection and servicing of fire alarm system

Page 1 of 2

Batteries	Standby battery disconnected and full load alarm simulated	☐ yes	☐ no
	Batteries and their connections examined and tested	☐ yes	☐ no
	Vented batteries examined to ensure the specific gravity of each cell is correct	☐ yes	☐ no
System checks	System checked by the operation of at least one detector or call point on each circuit	☐ yes	☐ no
	Devices used recorded in log book	☐ yes	☐ no
	Fire alarm sounder operation checked	☐ yes	☐ no
	All controls and visual indicators at control and indicating equipment checked	☐ yes	☐ no
	Operation of automatic transmission to an alarm receiving centre checked for both fire and fault signals	☐ yes	☐ no
	Ancillary functions of control and indicating equipment checked	☐ yes	☐ no
	Fault indicators and their circuits checked, where possible, by simulation of fault conditions	☐ yes	☐ no
	All printers checked for correct operation and sufficient quantities of consumables are present	☐ yes	☐ no
	Radio systems serviced in accordance with manufacturer's recommendations	☐ yes	☐ no
	All further checks and tests recommended by the manufacturer of the control and indicating equipment and other components of the system carried out	☐ yes	☐ no
Completion	On completion of the work,		
	▶ defects reported	☐ yes	☐ no
	▶ log book completed	☐ yes	☐ no
	▶ inspection and servicing certificate issued	☐ yes	☐ no
	Name and telephone number of maintenance organisation (if an external agency) prominently displayed at main control and indicating equipment?	☐ yes	☐ no

Test performed by

Name .... Sign .... Date ....

Date of next test

....

Page 2 of 2

Work to be carried out during each year on the fire alarm system		
Address of premises		
....		
Name and address of servicing organisation		
....		
....		
Category of system		....
Call points	Switch mechanism of every manual call point to be tested by	
	▹ removal of a frangible element, or	☐ yes
	▹ insertion of a test key, or	....
	▹ operation of the device as it would in the event of a fire	initials
Detectors	Detectors to be examined to ensure they are undamaged and unpainted	☐ yes
	Detectors to be functionally tested Heat source used for heat detectors (unless detector would be rendered inoperative) Smoke detectors checked with simulated smoke Optical beam detectors checked by introducing signal attenuation Aspirating fire detector systems functionally checked Carbon monoxide detectors functionally checked (Take care: CO is highly toxic) Flame detectors functionally tested Multisensor detectors tested	☐ yes .... initials
Sounders	Sounders and fire alarm devices checked for correct operation	☐ yes
	Visual devices are not obstructed from view and lenses are clean	☐ yes
Control & indicating equipment	Analogue values within range, where appropriate	☐ yes
	All unmonitored, permanently illuminated filament lamps replaced	☐ yes
	Radio signal strengths checked for adequacy, where appropriate	☐ yes
Cables	All readily accessible cable fixings are secure and undamaged	☐ yes
Program	Cause and effect programme correct	☐ yes
Power supply	Standby power supply capacity suitable	☐ yes
Other checks	All further annual checks and tests recommended by the manufacturer of the control and indicating equipment and other equipment carried out	☐ yes
Completion	On completion of the work,	
	▹ defects reported	☐ yes
	▹ certificate issued	☐ yes
Work performed by		
Name .... Sign .... Date		

▹ **Figure 8.4** Ongoing work to be carried out on a fire alarm system

Appointment of a new servicing organisation for a fire alarm system			
Address of premises			
Name and address of servicing organisation			
Category of system		....	
Call points	Adequate number of call points	☐ yes	☐ no
Detectors	Adequate number of detectors	☐ yes	☐ no
Sounders	Sounder levels from sounders comply with recommendations	☐ yes	☐ no
Power supplies	Standby power supply present (absence of a standby power supply must be highlighted to the responsible person)	☐ yes	☐ no
	Standby power supply(ies) meets recommendations	☐ yes	☐ no
Cabling	Cabling meets fire resistance recommendations	☐ yes	☐ no
Circuits	Monitoring of circuits meets recommendations	☐ yes	☐ no
Electrical safety	System meets the requirements of BS 7671	☐ yes	☐ no
False alarms	Numbers of false alarms within recommendations	☐ yes	☐ no
Premises	System suitable for premises (no changes in use, layout and construction of premises)	☐ yes	☐ no
Log book	Available	☐ yes	☐ no
Certificates	All certificates available	☐ yes	☐ no
Work performed by			
Name	Sign	Date	

▶ **Figure 8.5** Form to be filled in when appointing a new servicing organisation

Modification to a fire alarm system			
Address of premises			
Name and address of fire alarm maintenance, repair and modification organisation			
Details of modification			
Responsible person is aware of and agrees modifications to system		☐ yes	☐ no
Category of system		....	
Compliance	Proposed modifications do not detrimentally affect the compliance of the system with fire safety legislation	☐ yes	☐ no
Tests	All components, circuits, system operations and site specific software functions affected by the modification tested for correct operation	☐ yes	☐ no
	If one or more devices have been added to or removed from a circuit, one other device on the circuit should be tested	☐ yes	☐ no
	If the control and indicating equipment has been modified, at least one device on every circuit should be tested		
	If any additional load has been placed on the system:		
	▶ the rating of the power supply is adequate	☐ yes	☐ no
	▶ the capacity of the standby batteries is adequate	☐ yes	☐ no
	If software has been modified, other parts of the system must be tested to verify correct operation	☐ yes	☐ no
Drawings	As fitted drawings made available or existing drawings updated	☐ yes	☐ no
Certificate	Modification certificate issued	☐ yes	☐ no
Log book	Updated	☐ yes	☐ no
Work performed by Name .... Sign .... Date			

▶ **Figure 8.6** Record of modification to a fire alarm system

8.3 Dwellings

8.3.1 Fire detection and alarm systems for dwellings

In the UK around 80 per cent of all fire deaths and injuries occur in dwellings, a total of 450 to 500 deaths and 14 000 injuries per annum. The installation of a fire detection and alarm system can significantly reduce the risk of death or serious injury from fire. The fatality rate in fires in dwellings is three times higher where there is no smoke detector or where it is not working compared to dwellings where a fully functioning smoke detector is fitted. The installation of automatic fire detectors is, effectively, required in new dwellings to satisfy Building Regulations. Automatic fire detectors should be fitted in all dwellings.

In this document, discussions will be limited principally to those installed in dwellings, specifically Grade D, E and F systems.

8.3.2 System components

System components should correspond to the standards given in Table 8.9.

▶ **Table 8.9** System components for fire detection for dwellings

Component	Standard
Smoke alarms	BS 5446-1: 2000
Heat alarms	BS 5446-2
Smoke alarms systems intended to warn deaf or hard of hearing people	BS 5446-3
Manual call points	BS EN 54-11 for Type A (single action) manual call points
Point heat detectors	BS EN 54-5 for Class A1 or A2 detectors (ambient temperature below 40 °C)
Point smoke detectors	BS EN 54-7
Flame detectors	BS EN 54-10
Carbon monoxide fire detectors	In the absence of any international, European or British standard should meet the requirements of LPS 1265
Beam-type smoke detectors	BS EN 54-12
Optical smoke beam detectors	BS EN 54-12
Control and indicating equipment for Grade A systems	BS EN 54-2
Control and indicating equipment for Grade B systems	BS EN 54-2 or the recommendations of Annex C of BS 5839-6
Audible fire alarm devices for Grades A and B systems	BS EN 54-3
Power supply equipment for Grade A systems	BS EN 54-4
Cables	cables for Grade A and B systems should comply with the recommendations of BS 5839-1: 2002 Grade C systems should employ cables specified in BS 7671 that are suitable for the voltage and current. Grade D and E systems should employ normal domestic mains wiring cable if Grade F systems are to be interconnected, cable suitable for the voltage and current should be used

8.3.3 Grades of system

The grades of system are defined in Table 8.10.

▶ **Table 8.10** Grades of fire detection system

Grade	Description
A	a fire detection and fire alarm system which incorporates control and indicating equipment conforming to BS EN 54-2, and power supply equipment conforming to BS EN 54-4, and which is designed and installed in accordance with all the recommendations of sections 1 to 4 of BS 5839-1 except those in the following clauses for which the corresponding clauses of BS 5839-6 should be substituted BS 5839-1 BS 5839-6 16 13 18 14 20 18 25.4e) 15.2c) 27 21
B	a fire detection and fire alarm system comprising fire detectors (other than smoke and heat alarms), fire alarm sounders, and control and indicating equipment that conforms either to BS EN 54-2 (and power supply complying with BS EN 54-4) or to annex C of BS 5839-6
C	a system of fire detectors and alarm sounders (which may be combined in the form of smoke alarms) connected to a common power supply, comprising the normal mains and a standby supply, with central control equipment
D	a system of one or more mains-powered smoke alarms, each with an integral standby supply (the system may, in addition, incorporate one or more mains-powered heat alarms, each with an integral standby supply)
E	a system of one or more mains-powered smoke alarms with no standby supply (the system may, in addition, incorporate one or more heat alarms, with or without standby supplies)
F	a system of one or more battery-powered smoke alarms (the system may, in addition, incorporate one or more battery-powered heat alarms)

Grade A and B systems are systems of a type described in BS 5839-1.

In a Grade C system, the fire detectors are supplied with a common power supply unit with central control equipment, and this type of system normally incorporates a secondary rechargeable battery which is unlikely to be removed for other purposes.

A Grade D system consists of mains-powered smoke alarms with standby supplies. A battery or capacitor is provided to ensure protection is available under loss of mains conditions.

A Grade E system consists of mains-powered smoke and heat alarms and is potentially more reliable than battery-powered smoke alarms because it requires less attention by the user. The cost of the system is higher than battery-powered smoke alarms, as a mains supply and interlinking cables are required and the detectors themselves cost slightly more. Loss of mains results in loss of protection.

Grade F systems (battery-operated smoke and heat alarms) are the simplest form of fire detection and alarm system, are low cost and relatively simple to install. Smoke alarms to BS 5446-1 and heat alarms to BS 5446-2 give a low battery warning. A disadvantage is that removal of the battery disables the protection.

8.3.4 Categories of system

Fire alarm systems are usually installed in dwellings to protect life.

Fire alarm systems are divided into the following categories:

▶ **Table 8.11** Categories of fire alarm system

Category		Description and objectives
LD Objective of Category L systems is the protection of life (D means dwelling)	LD1	a system installed throughout the dwelling incorporating detectors in all circulation spaces that form part of the escape routes from the dwelling, and in all rooms and areas in which fire might start, other than toilets, bathrooms and shower rooms
	LD2	a system incorporating detectors in all circulation spaces that form part of the escape routes from the dwelling, and in all rooms or areas that present a high fire risk to occupants
	LD3	a system incorporating detectors in all circulation spaces that form part of the escape routes from the dwelling
PD Objective of Category P systems is the protection of property	PD1	a system installed throughout the dwelling incorporating detectors in all areas in which fire might start other than toilets, bathrooms and shower rooms
	PD2	a system incorporating detectors only in defined rooms or areas of the dwelling in which the risk of fire to property is judged to warrant their provision

Any statutory requirements imposed by enforcing authorities, and any requirements imposed by property insurers should state the Category of system required.

A system that is intended to fulfil the objectives of more than one Category of system needs to comply with the recommendations for each Category.

The Category of system needs to be defined in the specification and, except for Category LD1 or PD1 systems, the details of the areas of the building to be protected.

Class of dwelling	Minimum grade and category of system for installation in:					
	New or materially altered dwelling complying with BS 5588-1[a]		Existing dwelling Complying with BS 5588-1[a]		Existing dwelling Where fire precautions are of a lower standard than BS 5588-1[a]	
	Grade	Category	Grade	Category	Grade	Category

▶ **Table 8.12** Minimum Grade and Category of fire detection and alarm system for the protection of life

Single-family dwellings[b] and shared houses[c] with no floor greater than 200 m² in area

Class of dwelling	Grade	Category	Grade	Category	Grade	Category
Owner-occupied bungalow, flat or other single-storey unit	D	LD2[d]	F[e]	LD3[f]	D	LD2[g]
Rented bungalow, flat or other single-storey unit	D	LD2[d]	F[e, h]	LD3[f]	D	LD2[g]
Owner-occupied maisonette or two-storey house	D	LD2[d]	F[e, f]	LD3[f]	D	LD2[g]
Rented maisonette or two-storey house	D	LD2[d]	D[e, f]	LD3[f]	D	LD2[g]
Three-storey house	D	LD2[d]	D[e]	LD3[f]	D	LD2[g]
Four (or more)-storey house	B	LD2[d]	D	LD2[d, i]	B	LD2[g]

Single-family dwellings[b] and shared houses[g] with one or more floors greater than 200 m² in area

Class of dwelling	Grade	Category	Grade	Category	Grade	Category
Single-storey dwelling (bungalow, flat or other)	D	LD2[d]	D	LD3[f]	D	LD2[g]
Two-storey house (or maisonette)	B	LD2[d]	B	LD2[d, i]	B	LD2[d, g]
Three (or more)-storey house	Grade A, Category LD2, with detectors sited in accordance with the recommendations of BS 5839-1 for a Category L2[d, i, j] system					

Houses in multiple occupation[k] (HMO)

Class of dwelling	Grade	Category	Grade	Category	Grade	Category
Single or two storey with no floor greater than 200 m² in area	D	LD2[d]	D	LD3[l, f]	D	LD2[l, g]
All other HMOs: individual dwelling units within the HMO, comprising two or more rooms	D[m]	LD2[d]	D[m]	LD3[n]	D[m]	LD2[g]
communal areas of the HMO	Grade A, Category LD2, with detectors sited in accordance with the recommendations of BS 5839-1 for a Category L2 system[o]					

Notes:

a Guidance on the national *Building Regulations*, in England and Wales, is detailed in *Approved Document B* published by the Office of the Deputy Prime Minister. In Scotland, the *Technical Handbooks* are published by the Scottish Executive. In Northern Ireland, *Technical Booklet E* is published by the Department of Finance and Personnel.

b Single family dwellings include dwellings with long-term lodgers, but not boarding houses.

c Shared houses are houses shared by no more than six persons generally living in a similar manner to a single family.

d Except for housing providing NHS supported living in the community, heat detectors should be installed in every kitchen and the principal habitable room. Where more than one room might be used as the principal habitable room, a heat detector should be installed in each such room. The detector in the principal habitable room (but not the kitchen) may alternatively be a smoke or carbon monoxide fire detector. However, a heat detector is preferred in view of its lower potential for false alarms and the lesser need for maintenance.

e Grade E if there is any doubt regarding the ability of the occupier to replace batteries but Grade D if there is a likelihood of the electricity supply being disconnected because the occupier is unable to pay.

f Category LD2 if a risk assessment justifies the provision of additional detectors.

g Detectors should be of a type and be so located to compensate for the lower standard of structural fire precaution. Further detectors may be necessary.

h For rented properties, the batteries in smoke alarms should have an anticipated life of five years. Battery removal should require a special tool.

i Further detectors may be necessary if a risk assessment requires them.

j Refer to BS 5839-1 which recommends detectors are installed in escape routes and, generally, in rooms opening on to escape routes. However, detectors may be omitted from rooms opening directly onto staircase landings and opening on to escape corridors of 6 m or less in length.

k Other than houses with long-term lodgers and houses shared by no more than six persons living in a similar manner to a single family.

l Detectors should be installed in communal circulation routes and within any circulation spaces in individual dwelling units comprising two or more rooms.

m The detectors in individual dwellings may be incorporated within the system installed in communal areas.

n Category LD2 if a risk assessment justifies the provision of additional detectors.

o Heat detectors should be installed in every communal kitchen. Heat or smoke detectors should be installed in every communal lounge.

8.3.5 False alarms

False alarms are common in dwellings, and as a result householders may disable smoke alarms to stop false alarms. Deaths and serious injuries can result from fires in dwellings where smoke alarms have been disabled by householders due to frequent false alarms. False alarms are not just a nuisance; they are seriously detrimental to fire safety.

False alarms due to equipment failure, malicious intent or other causes are rare.

A small proportion of false alarms can be caused by dust within smoke detectors, and it is important users have a means of silencing such false alarms. A facility to disable or reduce the sensitivity of a smoke detector at times when a false alarm is likely (when cooking) can be of benefit.

False alarms can result from:

▶ toasting of bread
▶ fumes from cooking
▶ steam from bathrooms, shower rooms and kitchens
▶ tobacco smoke
▶ dust
▶ insects
▶ aerosol spray
▶ smoke from an outside bonfire
▶ burning off paint with a blowlamp
▶ candles
▶ incense
▶ high humidity.

Avoidance of false alarms can be achieved by the following means:

▶ smoke detectors in hallways and corridors into which kitchens open should, normally, be of the optical type
▶ optical smoke detectors should not be sited close to rooms from which steam may issue such as poorly ventilated bathrooms, shower rooms and certain kitchens
▶ detectors should not be installed in bathrooms or shower rooms
▶ if a detector is to be installed in a kitchen, only a heat detector should be installed
▶ an ionisation chamber smoke detector should be installed in preference to an optical smoke detector in a dusty room or area or one where dense tobacco smoke is likely to occur
▶ means should be provided for silencing short-term unwanted alarms.

Refer to BS 5839-6 for full details.

8.3.6 Sounders

Sounders should produce a frequency of sound in the range of 500 to 1000 Hz except for smoke and heat alarms, which should not exceed 3500 Hz.

The fire alarm warning should be clearly distinguishable from the sounds produced by other alarm systems in the building. All fire alarm sounders should produce a similar sound.

Smoke alarms should be interlinked such that when a fire is detected by any smoke or heat alarm, an audible alarm is given by all smoke alarms in the dwelling.

Within a bedroom, unless there is a fire alarm sounder that will give a signal whenever fire is detected anywhere in the dwelling, in all Category LD systems a sound level of at least 85 dB(A) should be measured. For people who are deaf or hard of hearing, a higher sound pressure level may be needed.

8.3.7 Power supplies

A list of systems and their appropriate power supplies is given in Table 8.13. Figure 8.7 shows the supply to a Grade E system.

▷ **Table 8.13** Power supplies

Grade A, B and C systems	refer to BS 5839-6
Grade D systems	the mains supply to smoke and heat alarms should be either a single independent circuit from the dwelling's main distribution board or a separately electrically protected regularly used local lighting circuit
	smoke and heat alarms should be interconnected and, in this case, must be supplied from the same circuit unless the method of interconnection is by radio transmission
Grade E systems	the mains supply to smoke and heat alarms should be a single dedicated independent circuit from the dwelling's main distribution board
	smoke and heat alarms should be interconnected and, in this case, must be supplied from the same circuit unless the method of interconnection is by radio transmission
	the circuit supplying the smoke and heat alarms should preferably not be protected by an RCD unless one is required for reasons of electrical safety, then either the RCD should serve only the circuit supplying the smoke or heat alarms or the RCD protection of the fire alarm system should operate independently of any RCD protection for circuits supplying socket-outlets or portable equipment
Grade F systems	the batteries of smoke alarms and any heat alarms should be capable of supplying the normal load, including the additional load from routine weekly testing, for at least one year before the battery fault warning is given
	at the point at which the battery fault warning commences, the batteries should have sufficient capacity to give a fire alarm warning signal for at least 4 min or, in the absence of a fire, a battery fault warning for at least 30 days

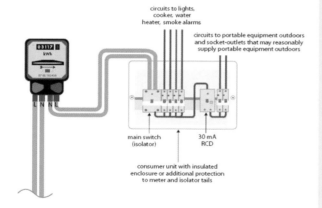

circuits to lights, cooker, water heater, smoke alarms

circuits to portable equipment outdoors and socket-outlets that may reasonably supply portable equipment outdoors

main switch (isolator)

30 mA RCD

consumer unit with insulated enclosure or additional protection to meter and isolator tails

▲ **Figure 8.7** Power supply to a Grade E system where the installation forms part of a TT system. The 100 mA time-delayed RCD provides protection for the fire alarm system (and other circuits) and operates independently of the RCD protection for the socket-outlets.

8.3.8 Cables

All cables should be selected and installed in accordance with the requirements of BS 7671 and the recommendations of BS 5839-6.

Grade A, B and C systems	refer to BS 5839-6
Grade D and Grade E systems	cables used for the mains supply to smoke alarms, any heat alarms and any interconnecting wiring may comprise any cable suitable for domestic mains wiring
	cables used for interconnecting smoke and heat alarms should be readily distinguishable from those supplying power (for example by colour coding)
	cables used for unmonitored circuits should be protected against damage
Grade F systems	cables suitable for the voltage or current are suitable
	cables used for unmonitored circuits should be protected against damage

▶ **Table 8.14** Wiring systems

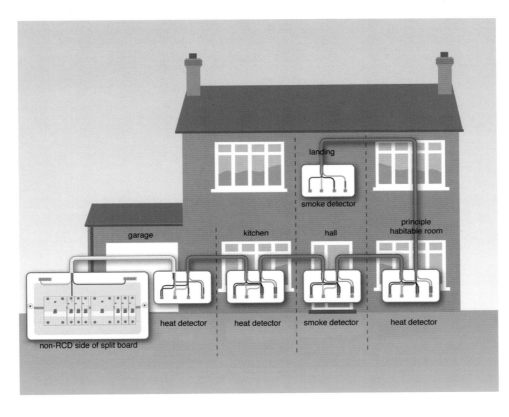

▶ **Figure 8.8** Installation of detectors and cabling in a two-storey house

Figure 8.9 Installation of detectors and cabling in a flat

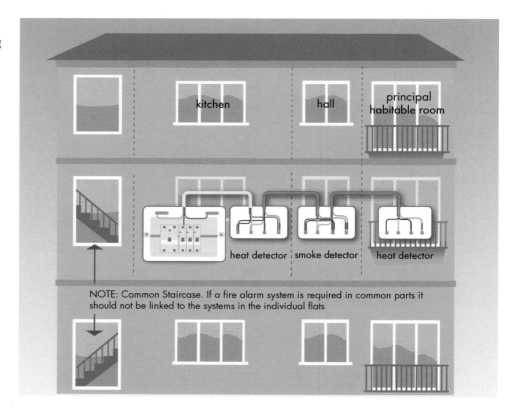

Figure 8.10 Installation of detectors and cabling in a bungalow

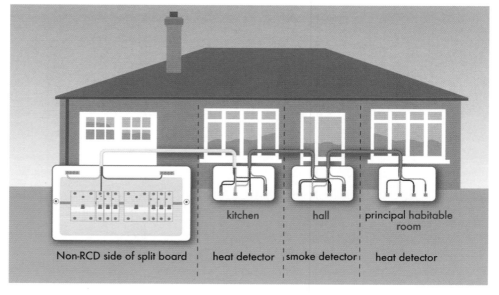

8.3.9 Installation, commissioning and certification

The installer should provide as-fitted drawings; installation should comply with the requirements of BS 7671. Cables should be neatly run and unnecessary joints or connections should be avoided wherever possible.

Other than MICC or SWA cables, cable should be given mechanical protection in any areas in which physical damage or rodent attack is likely. Where a cable passes through a wall, a small clearance hole should be provided. If additional mechanical protection is necessary, a smooth-bore sleeve should be sealed into the wall. The hole should be as small as reasonably practicable and made good with materials that ensure the fire resistance of the construction is not materially reduced.

Wiring systems penetrating walls, floors or ceilings should be internally sealed so as to maintain the fire resistance at the wall, floor or ceiling as well as being externally sealed to maintain the required fire resistance. Cables should be routed so that mechanical damage is avoided.

Tests made to the mains supply circuit should include earth continuity, polarity and earth fault loop impedance.

Insulation tests should be made of installed cables as required by BS 7671 – disconnect electronic equipment to avoid damage. Manufacturer's tests should be carried out.

At commissioning, the entire system should be inspected and tested to ensure that it operates satisfactorily and that, in particular:

▶ All manual call points and automatic fire detectors function correctly when functionally tested. Smoke detectors should be smoke tested with a simulated smoke aerosol that will not damage the detector. Heat detectors should be tested by means of a suitable heat source unless detector damage would otherwise result. The heat source should not have the ability to cause a fire; a live flame should not be used.
▶ Sounders should operate correctly, including any fire warning devices for people who are deaf or hard of hearing.

A certificate should have been issued to the user and this should be available for inspection.

In addition, for Grade F systems, sounders should be rigidly fixed to permanent construction, wiring between detectors should be installed and routed so that mechanical damage is avoided, each detector should be tested to demonstrate that it works and that any interlinking works and a certificate should be issued if installed by a professional installer.

8.3.10 User instructions
The supplier of the fire alarm system should provide the user with instructions which include:

▶ how to operate the system
▶ the importance of interlinking the detectors
▶ action in the event of an alarm
▶ how to avoid false alarms
▶ action in the event of a false alarm
▶ lifetime of components that are likely to require replacement
▶ routine testing of the system
▶ maintenance of the system
▶ need to keep a clear space around detectors
▶ special precautions relevant to any lithium batteries
▶ checking the system on reoccupation of the dwelling after a vacation
▶ need to avoid detector contamination by paint.

The operating instructions should be sufficient to enable a lay person to understand the system.

Silencing and disablement facilities should be explained but it should be stressed that system readiness must not be compromised.

Recommended action in the event of a fire must stress the importance of all occupants leaving the building as quickly as possible and that the fire service is summoned immediately regardless of the size of the fire.

Guidance should be given concerning the causes of false alarms and the precautions to be taken to avoid them.

Instructions for routine testing should be included. Instructions for maintenance should be included.

8.3.11 Routine testing

Instructions to users must stress the importance of routine testing. Systems should be tested weekly by pushing the test button.

If the dwelling has been unoccupied for a period during which the supply(ies) could have failed, the occupier should check that the system has not suffered total power failure and is still operable.

8.3.12 Maintenance

Grade A, B and C systems – refer to BS 5839-6.

Smoke alarms in Grade D, E and F systems should be cleaned periodically in accordance with the manufacturer's instructions.

Where experience shows that undue deposits of dust and dirt are likely to accumulate, so affecting the performance of the system before detectors are cleaned or changed, more frequent cleaning or changing should be carried out.

Model certificate for Grades B, C, D, E and F systems

Certificate of design, installation and commissioning* of the fire detection and fire alarm system at

....

.... Postcode ....

It is certified that the fire detection and fire alarm system at the above address complies with the recommendations of BS 5839-6 for design, installation and commissioning of a Category … Grade … system, other than in respect of the following variations

....

....

....

Brief description of areas protected (only applicable to Category LD2 and PD2 systems)

....

The entire system has been tested for satisfactory operation in accordance with the recommendations of clause 23.3p of BS 5839-6: 2004*

....

Instructions in accordance with the recommendations of clause 24 of BS 5839-6: 2004 have been supplied to:*

....

Test performed by

Name .... Sign .... Date ....

Date of next test ....

---

* Where design, installation and commissioning are not all the responsibility of a single organisation or person, the relevant words should be deleted. The signatory of the certificate should sign only as confirmation that the work for which they have been responsible complies with the relevant recommendations of BS 5839-6: 2004. A separate certificate(s) should then be issued for other work.

This certificate may be required by an authority responsible for enforcement of fire safety legislation, such as the building control authority or housing authority. The recipient of this certificate might rely on the certificate as evidence of compliance with legislation. Liability could arise on the part of any organisation or person that issues a certificate without due care in ensuring its validity.

▶ **Figure 8.11** Certificate for Grades B, C, D, E and F systems

Industrial and commercial switchgear

9

Where switchgear (both LV and HV) is installed, a copy of the Health and Safety Executive information document HSG 230 should be obtained. This document provides information for managers and technical staff concerning the risks that can arise from the use of high voltage and low voltage electrical distribution switchgear including oil-filled switchgear manufactured prior to 1970. Advice is given on precautions which should be taken to eliminate or control these risks.

9.1 Introduction and definitions

Switchgear is defined as a combination of one or more switching devices together with associated control, measuring, signal, protective, regulating etc. equipment completely assembled under the responsibility of the manufacturer with all the internal electrical and mechanical interconnections and structural parts.

▶ **Figure 9.1** *Keeping electrical switchgear safe*, HSE publication HSG 230

The three-phase electrical switchgear covered is in the range from 400 V up to 33 kV. The types covered embrace switchgear using sulphur hexafluoride (SF_6), vacuum, air and oil as the medium in which the interruption is made.

Switchgear falls principally into the following five types of device: isolator, circuit-breaker, switch, switch fuse and fuse switch (Table 9.1).

There are five types of operating mechanism used for these switching devices and they are defined in Table 9.2.

In general, switchgear has a proven record of reliability and performance. Failures are rare, but where they occur the results may be catastrophic. Tanks may rupture and, in the case of oil-filled switchgear, may result in the ejection of burning oil and gas clouds, causing death or serious injury to persons and major damage to plant and buildings in the vicinity of the failed equipment.

Modern switchgear using sulphur hexafluoride gas or a vacuum as the insulating medium has removed the hazard of vaporising and burning oil, but has introduced other risks that need to be managed.

Switchgear is generally located in substations and switchrooms, i.e. areas that are separated from the day-to-day activity of the premises and which, in many instances, are visited on a very infrequent basis. Such rooms are generally locked and access is usually restricted to authorised persons.

Accident experience has shown that failure usually occurs at, or shortly after, operation of the equipment. Thus, the way switchgear is operated, its condition and the

9

▶ **Table 9.1** Switching devices

	Isolator	Circuit-breaker	Switch	Switch fuse	Fuse switch[1]
	a switching device which is used to open (or close) a circuit when negligible current is interrupted or established or when no significant change in voltage across the terminals of each pole or phase of the isolator will result from the operation	a switching device capable of making and breaking, or closing and opening, a circuit under normal conditions and under abnormal conditions such as those of short-circuit	a switching device suitable for making or closing a circuit under normal and abnormal conditions such as those of short-circuit, and capable of breaking or opening a circuit under normal conditions	a switching device that is an integral assembly of switch and fuses in which a fuse is connected in series with a switch	a switching device in which a fuse link or fuse carrier constitutes the moving contact
Operating duties					
	an isolator has no rated making or breaking capability. An isolator may only be used for off-load or dead switching	a circuit-breaker is a fault make/fault break device	a switch is a fault make/load break device		
	The isolator, circuit-breaker and switch will safely carry their rated short-circuit current for their rated duration				

1 The electrical supply industry standards in the UK will not allow the supply of this type of equipment on any future orders

circumstances existing in the system at the time of operation, to a large extent, determine whether the equipment will safely perform its duty.

High voltage switchgear and equipment such as transformers require regular routine maintenance. This work will need to be entrusted to a competent organisation. This publication cannot provide all the information necessary with respect to high voltage maintenance, but only an outline of the work that is likely to be required is given to assist in placing orders for the work.

Generally, maintenance is considered as three activities: inspection, examination and overhaul. In the absence of manufacturer's recommendations, the recommended maximum periods between these maintenance activities are given in Table 9.3.

It is important to note that there may be circumstances where maintenance activities should be performed more frequently due to unfavourable conditions such as frequent operation of the switchgear or an onerous environment.

Dependent manual operation (DMO)	Indepedent manual operation	Dependent power operation	Independent power operation	Stored energy operation
an operation solely by means of directly applied manual energy such that the speed and force of the operation are dependent upon the action of the operator **DMO switchgear should be limited to off-load operation and must be replaced or modified as soon as possible**	a stored energy operation where the energy originates from manual power, stored and released in one continuous operation, such that the speed and force of the operation are independent of the action of the operator	an operation by means of energy other than manual, where the completion of the operation is dependent upon the continuity of the power supply (to solenoids, electric or pneumatic motors etc.)	a stored energy operation where the stored energy originates from an external power source and is released in one continuous operation, such that the speed and force of the operation are independent of the action of the operator	an operation by means of energy stored in the mechanism itself prior to the completion of the operation and sufficient to complete it under pre-determined conditions; this kind of operation may be subdivided according to **a** the manner of storing the energy **b** the origin of the energy **c** the manner of releasing the energy

▶ **Table 9.2** Operating mechanisms

It is important to note that all oil circuit-breakers must be overhauled as soon as possible after fault operation.

It would be good practice to examine the isolating contacts and external bushings of SF6 circuit-breakers as soon as practicable after fault operation.

Maintenance activity	Maximum period between maintenance activities
Inspection	1 year
Examination	5 years
Overhaul	15 years

▶ **Table 9.3** Frequency of maintenance for 11 000 V switchgear and 11 000/415 V transformers

9.2 Potential problems with switchgear

Switchgear can have the following potential problems (note that switchgear that is more than 25 years old can be particularly affected).

Persons not having knowledge of the equipment

Switchgear is usually operated by trained staff known as 'authorised persons', but due to changing employment patterns this may not be the case. Some organisations may have chosen to contract out all operational work and maintenance of their switchgear, and as a result there may be no-one in the organisation who understands the equipment, its safe operation (particularly when it is the dependent manually operated type) and either the need for maintenance or maintenance procedures.

The equipment being overstressed

Switchgear is described as being 'overstressed' when the potential fault energy of the electrical system (e.g. from a short-circuit) at the switchgear location exceeds the fault energy rating of the switchgear. When it is operated under fault conditions it is unable to cope with the resulting electrical and thermal stress, which can sometimes lead to catastrophic failure, i.e. total destruction of the switchgear. In the case of oil-filled switchgear, such failures are accompanied by burning gas clouds and oil mist which envelop anyone near the switchgear and have resulted in serious burn injuries or death.

Equipment has not been modified as per manufacturer's advice

Manufacturers have, over the years, issued details of modifications to existing equipment which should be carried out on switchgear to improve its safety. However, in some cases, because of changes in ownership, users may be unaware of the need to carry out these modifications. As a result the equipment may be incapable of performing its duty satisfactorily.

Equipment may have dependent manually operated (DMO) switchgear

This is where the movement of the contacts is directly dependent on the movement of the handle by the operator. The operating mechanisms of most switchgear, i.e. independent manual, dependent power, independent power and stored energy, do not in themselves result in any particular risks. However where switchgear is dependent manually operated (DMO), the operator closes or opens the switchgear by moving a lever or handle by hand. DMO levers/handles are fitted to both high and low voltage switchgear. These types of operating mechanism are no longer made.

Movement of the contacts in DMO switchgear is totally dependent upon the speed and actions of the person operating the levers/handles. Any hesitancy on the part of the operator is likely to lead to a serious and potentially fatal failure of the switchgear. For example, an operator may not realise that the circuit-breaker has not been completely closed and release the operating lever/handle, thus drawing an arc within the oil tank. The arc can result in catastrophic failure. It is essential that levers/handles of DMO switchgear are operated in a decisive and positive manner without any hesitation and as rapidly as possible, particularly over the latter portion of the closing operation. In addition, should a lever/handle be closed onto a system fault, the force needed is significantly greater than when closed in a normal system load current. In some cases it may be physically impossible to close (or open) the device under fault conditions; again this may result in failure. The risks resulting from the use of DMO switchgear that is overstressed are particularly high.

Such switchgear needs urgent replacement or modification.

Equipment may not be maintained properly

It is not unusual to find that switchgear has been neglected. Routine servicing such as oil changing, lubrication, contact refurbishment and verification of contact engagement has often not been carried out for many years. Deterioration due to corrosion may also have occurred. This is usually the result of oversight, lack of knowledge of the equipment, or pressures to keep the equipment, and hence the plant, in operation. In many cases the expertise in handling and maintenance techniques for insulating oil is lacking. Where oil-filled switchgear has been neglected, it is difficult to assess the actual fault capability of the switchgear in the state in which it is found.

Another common problem is that the armatures of trip and close coils may have been lubricated by personnel who were unaware that, in time, this oil would dry out leaving a sticky residue that is likely to cause a failure to trip or close.

Antireflex operating handles

High voltage oil-switches and isolators may be fitted with operating handles which are not antireflex type. A common cause of accidents/incidents with high voltage oil-switches and isolators is when an operator carries out an incorrect operation when moving the operating handle, e.g. switching from 'off' to 'earth' instead of from 'off' to 'on', and then immediately tries to reverse that incorrect operation, thus inducing a further fault. Any attempt to open these types of switching device when fault current is flowing is likely to lead to the operator being enveloped in a cloud of burning oil and oil vapour. Operators have been killed as a result. To combat this, many manufacturers have produced antireflex operating handles for their equipment. These handles are a one-way operating device and have to be removed and relocated before carrying out a further operation. This built-in time delay means that when the incorrect operation is reversed, no fault current is flowing (as the circuit protection will have operated to interrupt the current flow), and there is no likely failure of the switch. The built-in time delay is also important when closing from 'off' to 'on' onto a known fault.

9.3 Management of switchgear

Persons using or responsible for switchgear must have a management system in place that will ensure the equipment will operate safely when called upon. The management system should include the following:

a Policies and procedures covering the installation, commissioning, operation, maintenance and decommissioning of the equipment.
b An appropriate system of records.
c Definition of responsibilities.
d Training requirements and records.
e Procedures for auditing the effectiveness of the procedures.

9.4 Documents

Three documents must be available and up to date: a network diagram, an asset register and maintenance records.

Network diagram

A network diagram is a convenient way of displaying the interconnection of plant items and switchgear bearing in mind there may be several switchrooms and supplies on one site.

At each location the maximum prospective fault current must be determined and annotated on the diagram and compared to the fault level capabilities of the equipment at that point in the installation. It is preferable to request the fault level at the origin of the installation from the electricity distributor. The maximum prospective fault level will depend upon the configuration of the supply company's network. If fault levels are calculated, the network configuration giving the highest fault level must be presumed.

From this basic information, any potential risks such as overstressing can be identified. In some cases, sufficient technical expertise may not be available in-house to carry out an assessment of risk and to decide on the appropriate precautions. However, having identified that a problem exists or may exist, switchgear users should be able to reach decisions about seeking further help from suitably competent organisations including regional electricity companies, switchgear manufacturers, switchgear maintenance companies with particular expertise in older types of switchgear and consulting organisations specialising in switchgear.

Asset register

Before carrying out any work on industrial switchgear, including operation or routine maintenance, a register should be prepared to identify the manufacturer, model, type, maintenance and modification history for each item of equipment.

Note: the trade association, British Electrotechnical and Allied Manufacturers Association (BEAMA), may also be able to provide help and guidance as to other sources of information and expertise.

A competent person must establish whether the switchgear is in suitable order and, after routine maintenance, can be used. If this is not the case, it is then important to take precautions to reduce the risk, pending modification or replacement. If the switchgear does not have sufficient fault rating, all live operation and automatic tripping of the switchgear should be prohibited and prevented. All access to the switchgear when live should be prevented.

It is stressed here that failure of such switchgear can result in fatalities.

Maintenance records

A minimum requirement would be to record the date of the last maintenance (and oil change where applicable) and in the case of a circuit-breaker, the number of fault operations since it was last maintained. The visual examination performed after a fault should be recorded.

Note the statement at the end of 9.1 (above Table 9.3) regarding post-fault maintenance of oil-filled and SF_6 circuit-breakers.

9.5 Switchgear must be fit for purpose

With knowledge of the maximum prospective fault current at each location and the load at each location, it may be determined if the switchgear is potentially fit for the purpose and if, after maintenance and if necessary repair, it may be safely operated.

Figure 9.2 33 kV fixed pattern vacuum circuit-breaker

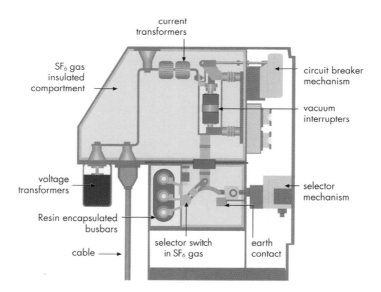

Asset register of switchgear

Switchgear owner ....

Address/location ....

	Item 1	Item 2	Item 3
Location			
Switch no.			
Manufacturer			
Type reference			
Type of equipment			
Serial no.			
Year of manufacture			
Date of installation			
Voltage rating			
Current rating			
Fault rating			
Is fault rating a certified or assessed value?			
Type of operating mechanism (dependent manual, independent manual, dependent power, independent power or stored energy)			
Details of modifications			
Details of repairs			
Date equipment last maintained or serviced			
For oil circuit-breakers, whether or not it is plain break (i.e. equipment without arc control devices)			
Type of electrical protection and detail of settings (e.g. HRC fuse, 30 A)			
Is switchgear fit for purpose?			
Remarks			
Date			
Completed by (name and signature)			

▶ **Figure 9.3** Switchgear asset register

All organisations with responsibility for switchgear should have available a record of their switchgear.

The fault current rating of a particular item of switchgear should be obtained from the equipment rating plate, or by enquiry of the manufacturer (the serial number will need to be quoted). If the fault rating of the switchgear does not meet that of the fault level, the switchgear must be replaced. If the switch is oil-filled, particular care must be taken.

9.6 Care and maintenance of oil-filled switchgear

Oil-filled switchgear has been made for 60 years and represents no risk if manufactured to current standards, maintained and operated correctly. However, should there be a fault on oil-filled switchgear there could be an explosion of burning oil, which provides not only a risk of fire to the premises but also a very real risk of death or serious injury to anyone who might be operating or close to the switchgear. For these reasons it is particularly important to carry out thorough investigations into the maintenance and modification history of any oil-filled switchgear installed. Responsible persons have obligations set out in the *Health and Safety at Work etc. Act*, the *Management of Health and Safety at Work Regulations* and the *Electricity at Work Regulations*.

The main failure modes for oil switchgear are:

a *Faults within oil compartments*. These are invariably catastrophic, with an explosion and fire often involving personal injury or fatality and serious damage to the building.
b *Failure of oil-filled circuit-breaker to trip due to mechanism or protection faults* usually results in an extended disconnection time due to the upstream circuit-breaker being called upon to trip. Failure to trip can also occur due to over oiling of the trip mechanism. See Section 9.2 for a discussion of potential problems with switchgear.
c *Solid insulation faults*. These are insulation faults occurring within the insulation itself or across its surface.

The major risk is that of catastrophic failure due to faults within the oil compartment. Such faults can result from:

▸ contaminated insulating oil
▸ poor maintenance of arc interruption system
▸ breakdown of solid insulation
▸ breaking fault current above rated capability of the device
▸ internal component failure.

Minimising the risk of catastrophic failure can be achieved by inspection and maintenance.

9.7 Inspection

A regular inspection is recommended. At the time of the inspection, remedial work should be prioritised. Figure 9.4 lists recommended inspections.

Inspection of substation, switchroom or switchgear (applicable for SF$_6$, vacuum, air or oil-filled switchgear)			
Switchgear owner			
Address/location			
Inspection of	Result	Remedial action	Recommendation code
Switchgear environment			
Check switchroom access and surrounds including fences and external walls if outdoors Verify no indications of trespass or interference			
Presence and legibility of warning notices			
Switchroom internal fabric			
Fire fighting equipment			
General housekeeping			
Signs of water ingress or dampness			
Access restricted to authorised persons			
Switchgear checks			
Check for high temperature in switchroom			
Presence of smoke			
Smell of 'hot' substances such as oil or compound			
Audible discharges or arcing			
Smell of ozone			
Signs of leaked oil in vicinity of oil circuit-breaker tank			
Signs of fresh compound leaks			
Distortion or evidence of sooting on enclosures			
Switchgear condition			
General condition of exposed busbars and air-break switches (where present)			
General condition of the switchgear (rust, oil leaks, oil level gauge, missing bolts)			
Compound leaks from cable boxes, busbar chambers, band joints and end caps			
Ammeters, voltmeters, operation indicators, protection equipment			
Labelling, padlocks and key exchange interlocks			
General condition			
Overall condition of substation, switchroom or switchgear subject to the limitations of a visual inspection			
Comments:			
Date			
Completed by (Name and signature)			
Note: recommendation code 1 means that urgent attention is required; users of the substation, switchroom or switchgear may be at risk recommendation code 2 means that improvement is required recommendation code 3 means that further investigation is required			

▷ **Figure 9.4** Recommended inspections for switchgear

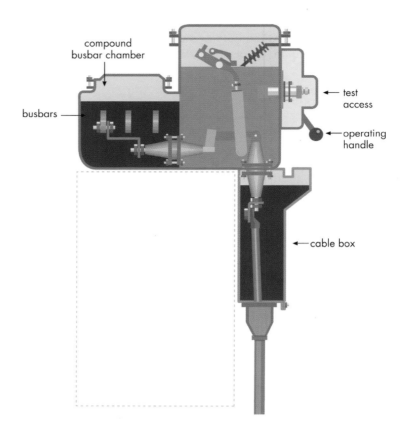

▶ **Figure 9.5**
Arrangement of an
11 kV switch

compound
busbar chamber

busbars

test
access

operating
handle

cable box

9.8 Maintenance

Maintenance must only be performed by trained, competent staff who are familiar with the equipment.

Caution: switching is required in order to release the equipment for maintenance, and personnel need to be aware that the risk of failure is greatest during a switching operation.

Caution: maintenance must be carried out by competent personnel. Errors can leave the equipment at greater risk of failure than if the maintenance had not been carried out.

Figures 9.6–9.9 offer a system for recording maintenance performed on oil-filled, SF_6, vacuum and air-break switchgear, respectively.

9.8.1 Oil-filled switchgear
Oil-filled switchgear requires regular time-based maintenance, and, provided such maintenance is performed, will prove to be extremely reliable. Recommendations from manufacturers or specialist organisations should be followed.

Detailed maintenance procedures can be found in BS 6423: 1983 and BS 6626: 1985.

Frequency of maintenance
Switchgear should be maintained at a frequency advised by the manufacturer or other competent organisation. Local conditions may dictate more frequent maintenance if outage due to switchgear malfunction is to be avoided. Switchgear that regularly interrupts large load currents will require more frequent maintenance attention.

Maintenance of oil-filled switchgear – external and internal examination

Switchgear owner ....

Address/location ....

Switchgear condition – external	Result and comments
General condition of exposed busbars and air-break switches (where present)	
General condition of the switchgear (rust, oil leaks, oil level gauge, missing bolts)	
Compound leaks from cable boxes, busbar chambers, band joints and end caps	
Ammeters, voltmeters, operation indicators, protection equipment	
Labelling, padlocks and key exchange interlocks	
Switchgear condition – internal	
Isolation procedures implemented	
Examination and cleaning of the tank interior, internal mechanism, contacts, arc control devices, bushings, phase barriers and tank lining	
Dressing, refurbishing or replacing main/arcing contacts (including contact alignment check using oil circuit-breaker slow-close facility)	
Cleaning of arc control devices or replacement if burnt or worn beyond acceptable tolerances (cross-jet pots, turbulators etc.)	
Replacement of insulating oil with new, reclaimed or reconditioned oil	
Insulating oil tested before introduced into equipment (see Section 9.12 – Further information)	
Lubrication of operating mechanism and adjustment where required	
Replacement of seals and gaskets, clearing vents and checking indicator windows	
Examination of primary isolating contacts for damage, burning, corrosion – cleaning and refurbishing (as necessary)	
Checking and lubrication of the oil circuit-breaker isolating mechanism	
Checking correct function of position indicators and interlocks	
Checking shutter operating mechanisms (as appropriate)	
Examining inside of cable termination chambers and current transformer chambers (as appropriate)	
Examination and checking of voltage transformer (as required)	
Secondary injection testing on circuit-breaker protection system (or, if this is not scheduled, carry out manual trip test)	
On fuse switches and switch fuses, trip testing with an appropriate fuse trip-testing device	
Examination of secondary contacts, wiring and auxiliary switches	
Check the truck goes fully into position and switchgear is level, as appropriate, when putting back into service	
Vents, if provided, are unobstructed and seals are intact	
Date	
Completed by (name and signature)	
Recommended date for next maintenance	
Note: During the maintenance of oil-filled switchgear, the tank cover should be removed for the minimum time necessary to reduce the risk of contamination of the tank interior by moisture, airborne particles, dust, insects and vegetation (if outdoors)	

▷ **Figure 9.6** Maintenance of oil-filled switchgear

Overstressed and/or DMO (Dependent Manually Operated) switchgear needs special attention. If such switchgear has not been maintained in the last three years, then maintenance should be carried out immediately and thereafter on a frequent basis.

Trip testing of circuit-breakers provides an operational test and exercises the mechanism. It can be carried out more frequently than the internal maintenance, within operational constraints. Annual trip-testing is considered satisfactory by many organisations. Trip testing may be performed to see if the circuit-breaker opens or it may be performed to establish that the breaker opens within the correct time. The latter would be preferable but, to be applied, would need routine test information from the manufacturer, together with timing equipment.

Insulating oil
Sampling of oil in service and analysis can provide valuable information on the state of switchgear. Further information is available in HSG 230 and BS 5730.

Cleaning and inspection of oil-filled chambers
Detailed information on the cleaning of oil-filled chambers and tanks is given in HSG 230. Danger can arise if:

▶ cloths, sponges or wipes are used which may leave fibres or particles of sponge in the chamber
▶ insulating oil is not clean or contains contaminants; contaminants can be deposited in the oil from switching operations; water in insulating oil can result in fungal growth
▶ personnel performing cleaning operations are not aware of the hazards from contaminants, inhalation of oil spray and ingestion; respirators must be worn at all times.

Post-fault maintenance
Oil-filled circuit-breakers that have closed onto a fault or interrupted a fault should be maintained as soon as possible and such maintenance must include:

▶ inspection and cleaning of all insulation within the tank to eliminate carbon, metal vapour and particle contamination
▶ restoration of the contacts and arc control devices to an acceptable condition, including a check on contact alignment by slow-closing the breaker
▶ replacement of the insulating oil with new, reclaimed or reconditioned oil
▶ inspection of the tank, tank gaskets and tank internal mechanism for signs of damage or distortion
▶ ensuring that external bushings and primary isolating contacts are in good order.

9.8.2 Care and maintenance of non-oil switchgear
Non-oil switchgear makes use of sulphur hexafluoride SF_6, vacuum or air as the interrupting medium. More modern designs employing SF_6 or a vacuum remove moisture or dust which can cause degradation of the interrupting medium, but nonetheless such switchgear requires maintenance. Two particular issues are:

1 With SF_6, problems will arise if gas is lost due to defective or worn seals.
2 With vacuum switchgear, X-rays may be generated when the gap is stressed at an excessively high voltage.

Note that this potential problem was recognised by manufacturers many years ago and interrupter designs were arranged to be such that X-rays will not be produced when normally recommended test voltages are applied. These test voltages are normally about three times the system voltage. The actual values used should be those specified by the manufacturer.

As with oil switchgear, inspection and maintenance will reduce the likelihood of catastrophic failure.

Inspection
The inspection form given in Figure 9.4 may be used when inspecting SF_6, vacuum or air switchgear.

Maintenance
Detailed guidance on the maintenance can be found in BS 6423: 1983 and BS 6626: 1985. Although SF6 and vacuum switchgear is designed to be low maintenance, maintenance is still required and maintenance is usually based on a time interval approach. Rigorous application of the manufacturer's advised time schedule should ensure high reliability.

9.8.3 Sulphur hexafluoride gas (SF$_6$)
Sulphur hexafluoride gas in its pure state is inert, colourless, non-flammable and non-toxic. However, like nitrogen, it will not support life and a large volume can cause suffocation. The gas is about five times heavier than air and will tend to accumulate at lower levels, for example in cable trenches and tunnels. The majority of modern switchgear up to 33 kV uses sealed containment with the gas at a small positive pressure (typically 0–1 bar). The equipment is factory-assembled and tested and no further handling of the gas is expected during the operating life of the switchgear.

Handling and safety precautions
Notices should be posted warning staff that SF_6 is being used in switchgear. In the unlikely event that the gas is to be handled, this is best left to suitably experienced and competent persons. Under no circumstances should the gas be released to the atmosphere. If the chamber holding the gas has experienced arcing, gas removal and overhaul should be sublet to a specialised experienced company, as metallic fluorides, which are the form of salts of acids, will almost certainly have been produced and specialised handling equipment will be needed. A policy should be agreed which identifies the experts who will be called in, in the unlikely event that it becomes necessary. SF_6 is a greenhouse gas and although its effect on global warming is likely to be small, control over its use is essential. The European electricity industries have agreed a set of actions with the manufacturers of the gas to reduce emissions to the atmosphere and recommend good housekeeping with the following objectives:

▸ SF_6 should not be deliberately released into the atmosphere
▸ SF_6 should be recycled and reused to the maximum extent possible
▸ losses of SF6 from electrical equipment should be minimised
▸ all new equipment should allow for recycling.

Procedures for safe handling are given in technical information such as *IEC Technical Report 1634, EA Engineering Recommendation G69.*

Maintenance of sulphur hexafluoride (SF$_6$) switchgear	
Switchgear owner	
Address/location	

Switchgear condition – external	Result and comments
Inspection of the external condition	
General condition of exposed busbars (where present)	
General condition of the switchgear (rust, missing bolts)	
Compound leaks from cable boxes, busbar chambers, band joints and end caps	
Ammeters, voltmeters, operation indicators, protection equipment	
Labelling, padlocks and key exchange interlocks	
Notice advising switchgear containing SF$_6$ has been posted	
Check gas pressure	
If topping up is necessary, refer to HSG 230 for details	
Switchgear condition – internal	
Isolation procedures implemented	
Inspection, adjustment and lubrication of mechanisms (including shutters where appropriate)	
On withdrawable equipment, examination of primary isolating contacts for damage, burning, corrosion – cleaning and refurbishment as necessary	
On withdrawable equipment, checking and lubricating of circuit-breaker isolating mechanism	
Checking correct function of position indicators and interlocks	
Examination of inside of cable termination chambers and other chambers as appropriate, removal of surface contamination from accessible solid insulation (where applicable)	
Examination and checking voltage transformer (as required)	
Secondary injection testing on circuit-breaker protection system (or, if this is not scheduled, carry out manual trip test)	
Examination of secondary contacts, wiring and auxiliary switches	
Date	
Completed by (name and signature)	
Recommended date for next maintenance	

▶ **Figure 9.7** Maintenance of SF$_6$ switchgear

Maintenance of vacuum switchgear

Switchgear owner ....

Address/location ....

Switchgear condition – external	Result and comments
Inspection of the external condition	
General condition of exposed busbars	
General condition of the switchgear (rust, missing bolts)	
Compound leaks from cable boxes, busbar chambers, band joints and end caps	
Ammeters, voltmeters, operation indicators, protection equipment	
Labelling, padlocks and key exchange interlocks	
Check vacuum pressure	
Switchgear condition – internal	
Isolation procedures implemented	
Measurement of contact wear where a measurement method is available	
Check on the vacuum integrity, e.g. by a high voltage pressure test (Caution X-rays may be emitted). Note that failure of a high voltage pressure test does not necessarily mean that the interrupter has lost vacuum.	
Inspection, adjustment and lubrication of mechanisms (including shutters where appropriate)	
On withdrawable equipment, examination of primary isolating contacts for damage, burning, corrosion – cleaning and refurbishment as necessary	
On withdrawable equipment, checking and lubricating of circuit-breaker isolating mechanism	
Checking correct function of position indicators and interlocks	
Examination of inside of cable termination chambers and other chambers as appropriate, removal of surface contamination from accessible solid insulation (where applicable)	
Examination and checking voltage transformer (as required)	
Secondary injection testing on circuit-breaker protection system (or, if this is not scheduled, carry out manual trip test)	
Examination of secondary contacts, wiring and auxiliary switches	
Date	
Completed by (name and signature)	
Recommended date for next maintenance	

▷ **Figure 9.8** Maintenance of vacuum switchgear

Maintenance of air-break switchgear	
Switchgear owner	
Address/location	
Switchgear condition – external	Result and comments
Inspection of the external condition	
General condition of exposed busbars	
General condition of the switchgear (rust, missing bolts)	
Compound leaks from cable boxes, busbar chambers, band joints and end caps	
Ammeters, voltmeters, operation indicators, protection equipment	
Labelling, padlocks and key exchange interlocks	
Check vacuum pressure	
Switchgear condition – internal	
Isolation procedures implemented	
Examination of the main/arcing contacts for excessive burning/damage – recondition or renew as required, taking account of manufacturer's requirements for different contact construction and materials	
Checking/adjusting spring contact force and contact alignment as required	
Removal, examination and cleaning of arc chutes – renew if damaged or eroded	
Inspection, adjustment and lubrication of mechanisms (including shutters where appropriate)	
On withdrawable equipment, examination of primary isolating contacts for damage, burning, corrosion – cleaning and refurbishment as necessary	
On withdrawable equipment, checking and lubricating of circuit-breaker isolating mechanism	
Checking correct function of position indicators and interlocks	
Examination of inside of cable termination chambers and other chambers as appropriate, removal of surface contamination from accessible solid insulation (where applicable)	
Examination and checking voltage transformer (as required)	
Secondary injection testing on circuit-breaker protection system (or, if this is not scheduled, carry out manual trip test)	
Examination of secondary contacts, wiring and auxiliary switches	
Date	
Completed by (name and signature)	
Recommended date for next maintenance	

▶ **Figure 9.9** Maintenance of air-break switchgear

9.9 Condition monitoring/assessment by partial discharge techniques

Partial discharge measurements provide valuable information on the condition of insulation in high voltage plant. Available non-intrusive techniques include measurement of transient earth voltages, ultrasonic detection and radio frequency interference detection.

Thermographic surveys involve infra-red detection using either thermal imaging or non-contact thermometers. Such surveys can be useful on open-fronted low-voltage switchgear only where live parts can be safely exposed and remotely scanned and can be used to detect overloaded conductors and connections and overheating bushings.

Mechanism timing tests can help to identify distortion or damage of the metal parts or binding in the mechanism resulting in slow opening or closing and eventually mechanism failure.

9.10 Refurbishment or replacement

Replacement of older switchgear can give the advantages that the new equipment will probably be smaller and opportunity will be available to modernise protection and control systems.

9.11 Maintenance schedules

In the absence of manufacturer's recommendations, typical maintenance schedules are shown for oil-filled circuit-breakers, switches, SF$_6$ switchgear and oil-filled transformers. A person placing an order for such maintenance should expect all these activities to be covered by inspection, examination and overhaul documentation.

Maintenance operations	Inspection (maximum of 1 year)	Examination (maximum of 5 years)	Overhaul (maximum of 15 years)
General inspection	✓	✓	✓
Inspect time limit fuses	✓	✓	✓
Check settings of protective relays	✓	✓	✓
Cleaning		✓	✓
Trip and closing mechanism		✓	✓
Contacts (main and arcing)		✓	✓
Trunk or truck winding mechanism		✓	✓
Shutters		✓	✓
Interlocks	✓	✓	✓
Bushings	✓	✓	✓
Auxiliary contacts		✓	✓
Secondary wiring and fuses		✓	✓
Insulation test		✓	✓
Earth connections		✓	✓
Heaters		✓	✓
Insulating oil		✓	✓
Main and arcing contacts		✓	✓
Arc control devices		✓	✓
Switchgear spouts			✓
Venting and gas seals			✓
Tank and tank linings			✓
Operational check	✓	✓	✓
Voltage and current transformers			✓

▶ **Table 9.4** Oil-filled circuit-breaker maintenance schedule

▶ **Table 9.5** Oil-filled switch maintenance schedule

Maintenance operations	Inspection (maximum of 1 year)	Examination (maximum of 5 years)	Overhaul (maximum of 15 years)
General inspection	✓	✓	✓
Cleaning*		✓	✓
Withdrawal locking mechanism*	✓	✓	✓
Safety shutters*	✓	✓	✓
Insulators/bushings*			✓
Insulating oil			✓
Isolating contacts*			✓
Operating mechanism			✓
Main and arcing contacts			✓
Arc control devices			✓
High voltage fuse connections			✓
Switchgear spouts			✓
Tank and tank linings			✓
Interlocks	✓		✓
Auxiliary contacts			✓
Earth connections			✓
Weather shields			✓
Insulation test			✓
Operational check		✓	✓

* Withdrawable types only

▶ **Table 9.6** Sulphur hexafluoride (SF$_6$) switchgear maintenance schedule

Maintenance operations	Inspection (maximum of 1 year)	Examination (maximum of 5 years)	Overhaul (maximum of 15 years)
General inspection	✓	✓	✓
Cleaning		✓	✓
Leak tests (specialist skills needed)		✓	✓
Lubrication		✓	✓
Operational check	✓	✓	✓

▶ **Table 9.7** Oil-filled transformer maintenance schedule

Maintenance operations	Inspection (maximum of 1 year)	Examination (maximum of 5 years)	Overhaul (maximum of 15 years)
General inspection of exterior	✓	✓	✓
Oil level	✓	✓	✓
Breather	✓	✓	✓
Sampling and testing of suspect oil only	✓		
Sampling and testing of oil as fixed routines		✓	✓
Clean exterior			✓
Conservator, clean and inspect			✓
Check external connections and conductors			✓
Main tank			✓
Visual inspection of transformer core			✓
Check LV and HV connections			✓

9.12 Further information

The information included in this section is based on the HSE publication HSG 230 entitled: *Keeping electrical switchgear safe*. Information on distribution switchgear may be found in the book *Distribution Switchgear* by Stan Stewart, ISBN 0-85269-107-3 available from the IEE.

Further information may be found in: *IEC Technical Report 1634*, EA *Engineering Recommendation G69* and the following British Standards:

BS 5207: 1975 *Specification for sulphur hexafluoride for electrical equipment*

BS 5209: 1975 *Code of practice for the testing of sulphur hexafluoride taken from electrical equipment*

BS 5730: 2001 *Monitoring and maintenance guide for mineral insulating oils in electrical equipment*

BS 6423: 1983 *Code of practice for maintenance of electrical switchgear and controlgear for voltages up to and including 1 kV*

BS 6626: 1985 *Code of practice for maintenance of electrical switchgear and controlgear for voltages above I kV and up to and including 36 kV*

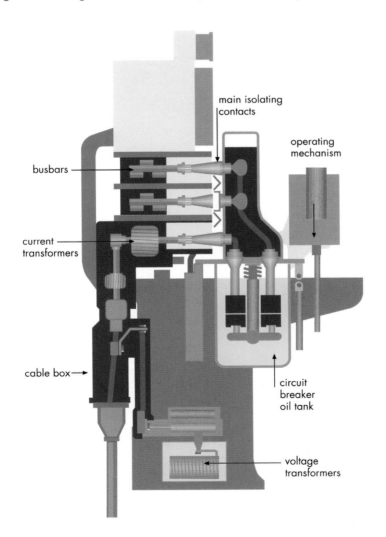

▶ **Figure 9.10** Horizontal isolation duplicate busbar

© The Institution of Engineering and Technology

Electromagnetic compatibility

10.1 Introduction

Any change of electric current will cause a change in the strength of its associated electromagnetic field which in turn can induce voltages and currents in other conductors in the field. The value of the induced voltage and current is dependent upon the rate of change of current (di/dt) in the conductor causing the electromagnetic changes. In the case of alternating current, the sinusoidal changes in the current can induce voltages and currents in other conductors. In addition, voltages and currents can appear in an installation due to lightning, switching, motor starting and faults on the installation or the distribution network. Such voltages and currents come within the definition of electromagnetic interference (EMI) and can disturb or damage information technology (IT) systems as well as other equipment with electronic components.

Low voltage installations may require protection against temporary overvoltages and faults between high voltage systems and earth, overvoltages of atmospheric origin or due to switching and voltage disturbances and electromagnetic disturbances.

Examples of where such effects can occur include:

▶ where large metal loops exist
▶ where different electrical wiring systems, such as power and IT cables, are installed in common routes
▶ near power cables carrying large currents with a high rate of rise of current (di/dt) (e.g. the starting current of lifts or currents controlled by rectifiers)
▶ in or near rooms for medical use where the electric or magnetic fields of electrical installations can interfere with medical electrical equipment.

10.2 Equipment that may be a source of electromagnetic emission

Electrical equipment sensitive to electromagnetic influences should not be located close to potential sources of electromagnetic emission, such as:

▶ devices switching inductive loads
▶ electric motors, in particular frequency-controlled motor drive systems
▶ fluorescent lighting
▶ welding sets
▶ computers
▶ rectifiers
▶ switched mode power supplies
▶ frequency converters or regulators
▶ lifts

> transformers
> switchgear
> power distribution busbars.

10.3 Reduction of electromagnetic effects

Many precautions can be taken to reduce electromagnetic effects, and these generally are best introduced at the design stage. As a maintenance activity, after the initial installation, it can be more difficult to reduce electromagnetic interference.

Guidance is given on reducing electromagnetic interference in installations of buildings in IEC standard IEC 60364-4-44: *Protection against voltage disturbances and electromagnetic disturbances.*

The general rules for avoiding EMI are:

a Equipment should meet the emission and immunity requirements of its standard. In the absence of any reference to electromagnetic capability (EMC) requirements within the equipment standard, the equipment should meet the emission and immunity requirements of EN 50081 or EN 50082.

b Electrical equipment that is sensitive to EMI should be fitted with surge protection devices and filters to reduce its susceptibility to mains-borne electrical interference.

c Metal sheaths of cables and metallic cable management systems should be bonded to a common bonding network.
    A bonding network (see Section 10.5) is a set of interconnected conductive elements that provide an electromagnetic shield for electronic systems at frequencies from direct current (d.c.) to low radio frequency (RF).

d Inductive loops should be avoided by selecting a common route in the building for power, signal and data circuits.

e Power and signal cables should be kept separate and should cross each other at right angles.

f Screened and/or twisted pair data cables should be used. The principle of the twisted pair is that the current in one conductor is matched by the return current in the other conductor. Similarly, changes in the electromagnetic field caused by other equipment induce equal and opposite voltages in each conductor of the pair, cancelling out the effect at the equipment.

g Potential sources of interference should be kept separate from sensitive equipment.

h Sensitive equipment such as IT communication cables should be kept separate from power cables subject to rapid changes of current, e.g. lift motor supplies and frequency-controlled motor drives.

i When using signal and data cables, manufacturer's instructions with respect to EMC requirements should be followed.

j Where a lightning protection system (LPS) is installed:
> the power, signal and data cables in the building should be separated from the down conductors of the LPS by a minimum distance which should be determined by the designer of the LPS in accordance with IEC 62305-3, or suitable screening should be provided
> metal sheaths or shields of power, signal and data cables should be bonded in accordance with the requirements for lightning protection given in IEC 62305-3 and IEC 62305-4.

k Equipotential bonding conductors and connections should have as low an impedance as possible by being as short as possible. In addition, the installation designer should consider increasing the cross-sectional area above that required for electrical safety in order to reduce the impedance of such conductors.

l Where an earthing busbar is employed to support the equipotential bonding system of a large information technology installation in a building, it may be installed as a closed ring. Note: This measure is preferably applied in buildings associated with the telecommunications industry

m Any metal enclosure, pipe or other metalwork should be bonded.

n A metallic screen should be provided around sensitive equipment.

o Single core conductors should be enclosed in earthed metal enclosures or equipment such as metallic conduit, trunking, ducting, tray and ladder systems.

p All metal service pipes and cables should enter the building at the same place, where possible.

q Fibre-optic or non-conducting links should be considered for communication networks between systems or buildings with separate earthing arrangements such as metal-free fibre-optic cables or other non-conducting systems for signal and data transmission, e.g. microwave signal transformer for isolation in accordance with IEC 61558 parts 2-1, 2-4, 2-6, 2-15 and IEC 60950-1.

r TN-C distribution networks can cause high levels of EMI and should be avoided.

s All the conductors of a particular system or circuit should be kept in very close proximity. In practice, the circuit could include phase conductors, neutral conductors and both protective and functional conductors. There may be difficulties in keeping all the conductors in close proximity. This is particularly true of protective conductors, where there may be parallel paths and loops.

t The use of a clean earth can be beneficial, i.e. a protective conductor connecting sensitive equipment directly to the main earth. Functional earth conductors should be run in proximity with the phase and neutral conductors supplying the equipment for the reasons explained above. Functional earth conductors can be particularly beneficial if the protective conductor is liable to pick up induced voltages from other equipment. Loops in protective conductors will naturally act as aerials picking up radiated field changes. Figure 10.1 illustrates measures that could be undertaken to protect against EMI in an existing building.

u Where there are problems in existing building installations due to electromagnetic influences the following measures may improve the situation (see Figure 10.1):
  ▸ use of fibre optic links for signal and data circuits
  ▸ use of Class II equipment
  ▸ use of double wound transformers in compliance with IEC 61558-2-1, IEC 61558-2-4, IEC 61558-2-6 or IEC 61558-2-15; the secondary circuit should preferably be connected as a TN-S system but an IT system may be used where required for specific applications.

v Unwanted tripping of circuits should be avoided by careful selection of device characteristics including, if necessary, time delayed characteristics.

w All earth electrodes associated with a building, such as protective, functional and lightning protection, should be interconnected.

10.4 Measures that can be applied for an existing building

1 cables and metal service pipes enter the building at the same place

2 inductive loops should be avoided by selecting a common route in the building for power, signal and data circuits

3 bonding connections should be as short as possible

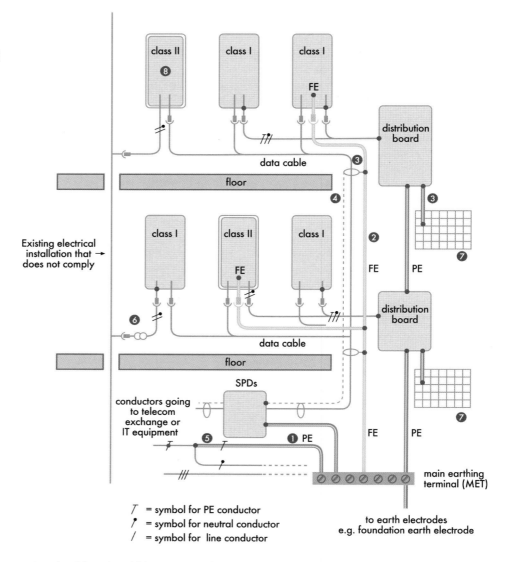

Figure 10.1 Measures to reduce EMI that can be applied to an existing building

4 signal cables should be screened or twisted pairs should be used
5 TN-C must not be used within the building
6 double-wound transformers may need to be used
7 a local horizontal bonding system can be employed (see IEC 60364-4-44)
8 use of Class II equipment

10.5 Bonding networks

There are different arrangements of bonding networks that are designed to reduce EMI, and Figure 10.2 illustrates a common meshed star bonding network which may be desirable in installations with a large amount of communication equipment. The mesh size should not exceed two metres square. Refer to IEC 60364-4-44 for full details.

10.6 Parallel Earthing Conductor (PEC)

Definition: a bypass equipotential bonding conductor or parallel earthing conductor is an earthing conductor connected in parallel with the screens of a signal or a data cable in order to limit the current flowing through the screen.

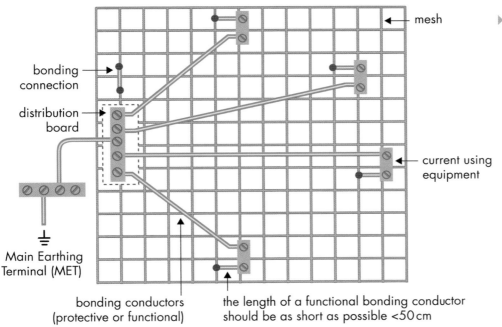

▶ **Figure 10.2** Common meshed star bonding network

mesh

bonding connection

distribution board

current using equipment

Main Earthing Terminal (MET)

bonding conductors (protective or functional)

the length of a functional bonding conductor should be as short as possible <50cm

The purpose of a parallel earthing conductor is to reduce the common mode current through leads that carry differential mode signals by reducing common impedance and areas of loops.

The parallel earthing conductor must be designed to withstand large currents when it is used as lightning protection, or as a power fault current return. Where a cable screen acts as a parallel earthing conductor, it is not designed to cope with such large currents. One approach is to route the cable through metallic construction elements, or conduits, which then act as a parallel earthing conductor for the total cable. Alternatively, a large cross-sectional area parallel earthing conductor can be connected in parallel with the screened cable and follow the same route with the cable screen and parallel earthing conductor being connected at both ends to the local earth of the equipment or apparatus.

For very long distances, additional connections of the parallel earthing conductor to the earthing system should be implemented at irregular intervals between the apparatus. These extra connections provide an early return path for the disturbance current through the parallel earthing conductor. For U-shaped conduits, shields and tubes, the additional earth connections should be made at the outside, preserving the separation from the inside ('shielding' effect).

Non-metallic cable management systems

Where an item of equipment is connected to the cabling system by unscreened cables, to improve EMC performance of non-metallic cable management systems, a single lead, as a parallel earth conductor, connected to the local earthing system at both ends, should be added inside the cable management system. The connections should be made on a low impedance metallic part (e.g. a large metal wall of the apparatus cabinet).

The parallel earth conductor should be designed to withstand large common mode and power fault currents.

10.7 Functional earthing conductor

Some electronic equipment requires a reference earth connection in order to function correctly. For such equipment, a functional earthing conductor is provided which connects from the equipment to a suitable means of earthing. Conductors for functional earthing may be metallic strips, flat braids or cables with circular cross-section. For equipment operating at high frequencies, metallic strips or flat braids are preferred and the connections should be kept as short as possible.

The colour cream should be used for functional earthing conductors. The colours green-and-yellow specified for protective conductors must not be used.

For equipment operating at low frequencies, the cross-sectional area of the functional earthing conductor should be the same as that required for the circuit protective conductor.

10.8 Commercial or industrial buildings containing significant amounts of IT equipment

The following additional specifications are intended to reduce the influence of EMI on IT. In severe cases, a common meshed bonding star network should be employed.

Sizing and installation of bonding ring network conductors

Equipotential bonding designed as a bonding ring network should have the following minimum dimensions:

- flat copper cross-section: 30 mm × 2 mm
- round copper diameter: 8 mm.

Parts to be connected to the equipotential bonding network

The following parts should be connected to the equipotential bonding network:

- conductive screens
- conductive sheaths or armouring of data transmission cables or of IT equipment
- functional earthing conductors.

10.9 Earthing arrangements and equipotential bonding of IT equipment for functional purposes

Earthing busbar

Where an earthing busbar is required for functional purposes, the Main Earthing Terminal (MET) of the building may be extended by using an earthing busbar. This enables information technology installations to be connected to the MET by the shortest practicable route from any point in the building. Where the earthing busbar is erected to support the equipotential bonding network of a significant amount of information technology equipment in a building, it may be installed as a bonding ring network.

The earthing busbar may be bare or insulated. The earthing busbar should preferably be installed so that it is accessible throughout its length, e.g. on the surface of trunking. To prevent corrosion, it may be necessary to protect bare conductors at supports and where they pass through out-walls.

Cross-sectional area of the earthing busbar

The effectiveness of the earthing busbar depends on the routing and the impedance of the conductor employed. For installations connected to a supply having a capacity of

200 A per phase or more, the cross-sectional area of the earthing busbar should be not less than 50 mm² copper.

10.10 Segregation of circuits

Information technology cables and power supply cables, which share the same cable management system or the same route, should be installed with a minimum separation to avoid disturbance. The minimum separation is related to many factors such as:

▶ the immunity level of equipment connected to the information technology cabling system to different electromagnetic disturbances (transients, lightning pulses, bursts, ring waves, continuous waves etc.)
▶ the way in which the equipment is connected to the earthing system
▶ the local electromagnetic environment (simultaneous appearance of disturbances, e.g. harmonics plus bursts plus continuous wave)
▶ the electromagnetic spectrum
▶ the distances that the cables run in parallel (coupling zone)
▶ the type of cable
▶ the coupling attenuation of the cables
▶ the quality of the attachment between the connectors and the cable
▶ the type and the construction of the cable management system.

Lightning protection installations 11

11.1 Design and installation

British Standard 6651: 1999 *Code of Practice for Protection of Structures against Lightning* provides comprehensive advice on the protection of structures against lightning, including:

▶ estimation of the need for protection
▶ system design
▶ inspection and testing
▶ records.

This publication considers periodic inspection and testing and the records that should be kept. The advice provided in this chapter is based on that given in BS 6651.

▶ **Figure 11.1** Lightning

If the original drawings of the installation are not available, it is necessary to take specialist advice as to the design of the system, as it may not be immediately apparent how the protection is provided. For example, in a steel frame structure the steel frame members may be acting as down conductors and their incorporation in the foundations of the building might lead to unacceptably high resistance without additional earth electrodes.

11.2 Inspection

A lightning protection system should be visually inspected by a competent person:

▶ during installation
▶ after completion
▶ after alteration or extension
▶ at fixed intervals, preferably not exceeding 12 months
▶ following a known lightning strike

in order to verify that it complies with the recommendations of BS 6651 and has not suffered damage or deterioration.

Inspection or testing should not be performed in adverse weather conditions when there is possibility of a lightning strike.

The inspection should confirm that the installation is installed as per the record drawings and is complete in all respects.

The mechanical condition of all conductors, bonds, joints and earth electrodes (including reference electrodes) should be checked and the inspector should check for evidence of corrosion or conditions likely to lead to deterioration. Observations should be noted.

If it is temporarily not possible to inspect some connections, for example due to site works, this should be noted. Any bonding of recently added services should be both inspected and recorded.

All changes, alterations or additions to the building structure which may require changes to the lightning protection system, including change of use, particularly such changes as the erection of masts, aerials or chimneys, must be recorded and brought to the attention of the responsible person.

It is a recommendation of BS 6651 (clause 16.1) that each down conductor of the installation be connected to an earth electrode.

11.3 Testing

Testing is an integral part of maintenance and should be carried out to confirm the visual inspection. Testing also confirms that there is continuity, and that the resistance to earth is within the limits required by the British standard. The results should be compared with the results from the previous tests and any substantial changes investigated. Environmental conditions should be recorded when carrying out tests. Resistance values may be higher during the height of summer when the ground is dry or in winter conditions when the ground is frozen. Measurements should, ideally, be taken under the worst conditions to be expected.

The following tests should be performed on the completion of the installation of a lightning protection system and at fixed intervals preferably not exceeding 12 months. The results should be recorded in a lightning protection system log book.

a The resistance to earth of each local earth electrode. Each local earth electrode will need to be disconnected at the test point between the down conductor and the earth electrode connection to perform this measurement (this is the 'isolated measurement').

Prior to disconnecting an earth electrode, it will need to be tested to ensure that there is no potential that could pose a risk of electric shock. The down conductor and the earth electrode should be checked for voltage both before and after disconnection. The voltmeter should be checked before and after making measurements with a proving unit.

Significant differences between the measurements for individual electrodes should be investigated.
b The resistance to earth of the complete earth termination system ('combined measurement'). This value must be measured when not bonded to other services.

The resistance to earth of the complete installation, comprising all the electrodes, should not exceed 10 Ω. If the resistance to earth of a lightning protection system exceeds 10 Ω, the value should be reduced except for structures on rock. If the resistance is less than 10 Ω but significantly higher than the previous reading, the cause should be investigated and any necessary remedial action taken. It is important to reduce the resistance to earth to 10 Ω or less as this reduces potential gradients around the earth electrodes when lightning currents are discharged. It may also reduce the risk of side flashing.

As previously stated, each down conductor of the installation should be connected to an earth electrode and the resistance to earth of each of these earth electrodes should not exceed the resistance value given below:

maximum value of electrode resistance ≤ 10 × No. of earth electrodes

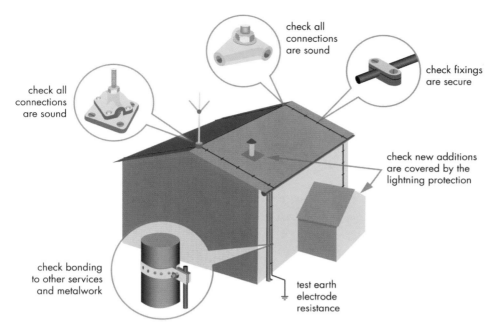

check all connections are sound

check all connections are sound

check fixings are secure

check new additions are covered by the lightning protection

check bonding to other services and metalwork

test earth electrode resistance

▶ **Figure 11.2** Checks to be made on a lightning protection system

For example, if there are ten electrodes in the installation, the resistance of each individual electrode should not exceed 100 Ω and if there are 20 electrodes in an installation, the resistance of each individual electrode should not exceed 200 Ω.

c A visual check on the condition of the conductors, bonds and joints or their measured electrical continuity.

The recommended method of testing is given in BS 7430.

11.4 Records

The following records should be kept on site by the person responsible for the lightning protection system and made available to engineers performing maintenance:

a Scale drawings of the installation showing the nature, dimensions, materials and positions of all component parts of the system. The drawings should be up-to-date and show any additions or changes made.

b The nature of the soil, in particular the presumed soil resistivity (Ω.m).

c Any special earthing arrangements.

d The type and position of the earth electrodes including the reference electrodes.

e The test conditions and results obtained.

f Any alterations, additions or repairs to the system.

g The results of all previous tests, including environmental conditions at the time of testing.

h The name of the person responsible for the installation or its upkeep.

i The presence of a label at the origin of the electrical installation.

> **This structure is provided with a lightning protection system that is in accordance with BS 6651 and the bonding to other services and the main equipotential bonding should be maintained accordingly.**

▶ **Figure 11.3** Lightning protection label

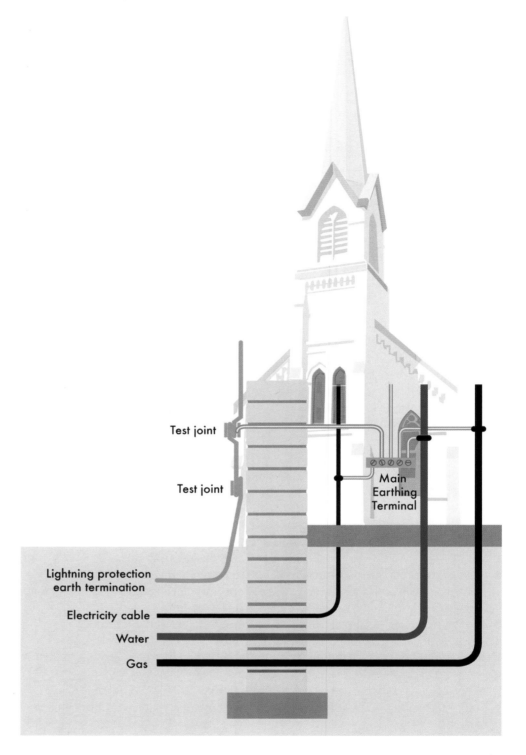

Test joint

Test joint

Main
Earthing
Terminal

Lightning protection
earth termination

Electricity cable

Water

Gas

11.5 Requirements placed by BS 7671

Lightning protection of buildings is excluded from the scope of BS 7671: 2001 *Requirements for Electrical Installations* by indent ix of Regulation 110-02.

However, the Regulations do require main equipotential bonding conductors complying with Section 547 to be connected between the Main Earthing Terminal of the electrical installation and, amongst other things, the lightning protection system, where such a system is installed (Regulation 413-02-02). Section 547 gives requirements relating to the required cross-sectional areas of such bonding conductors. In addition, where there is a lightning protection system, due account must be taken of the recommendations of BS 6651 (Regulation 541-01-03).

Where main bonding connections are to be made to a lightning protection system, it is advisable to consult the lightning protection system designer to establish the exact position of the bonding connection(s) due to the complexity of the recommendations of BS 6651.

Where, as is usually the case, the down conductors of a lightning protection system are connected together, only one main equipotential bonding connection is needed. The connection should normally be from the down conductor closest to the main earthing terminal of the electrical installation by the most direct route available. Figure 30 of BS 6651; 1999, reproduced above as Figure 11.4, includes a typical arrangement for such a connection. As a general rule each arrangement should be considered on its own merits and discussed with the authorities concerned.

The cross-sectional area of a main equipotential bonding conductor connecting a lightning protection system to the main earthing terminal of an electrical installation is to be not less than that required by Regulation 547-02-01 of BS 7671.

The bonding connection to a lightning protection system is normally made outdoors and may involve conductors of different shapes and dissimilar metals. It must be ensured that the electrical connection provides durable electrical continuity, adequate mechanical strength and includes measures to avoid corrosion (Regulations 526-01-01 and 522-05 refer).

> **Figure 11.5** A label in accordance with BS 951.

Safety electrical connection – do not remove

A permanent, durable label to BS 951 with the words 'Safety electrical connection – do not remove' should be affixed in a visible position at or near the connection of the main equipotential bonding conductor to the down conductor of the lightning protection system.

Exceptionally, lightning protection system designers may advise that a main bonding connection should not be made to a lightning protection system. In such circumstances, the consequent departure from Regulation 413-02-02 must be recorded on the Electrical Installation Certificate or Periodic Inspection Report and brought to the customer's attention. This is permissible only where the electrical contractor has requested the main bonding connection be made and has the lightning protection system designer's written objection to the connection being made on the grounds of safety.

11.6 Lightning protection system check sheet

A sample lightning system check sheet is given in Figure 11.6.

Lightning protection system check sheet

Address of premises that is fitted with the lightning protection system

....
.... Postcode

....

Responsible person

Date of inspection ....

Type of inspection   ☐ New installation
  ☐ Periodic inspection
  ☐ Following a lightning strike
  ☐ Following an addition or alteration

Weather conditions

....

Inspections

	yes	no	comments
Nature of soil			
Drawings available	☐	☐	....
Drawings up to date	☐	☐	....
Any new structures nearby (masts, aerials, buildings etc.)	☐	☐	....
Evidence of alterations or additions to system	☐	☐	....
Each down conductor connected to an earth electrode	☐	☐	....
Down conductors in satisfactory condition	☐	☐	....
Down conductor connections satisfactory	☐	☐	....
Evidence of corrosion	☐	☐	....
Evidence of damage	☐	☐	....
Label fitted at origin of installation	☐	☐	....
Main bonding connection provided	☐	☐	....
Main bonding connection satisfactory	☐	☐	....

▶ **Figure 11.6** Lightning protection system check sheet   *Page 1 of 2*

Test results

	Location	Condition	Resistance to earth Ω
Electrode no. 1	....	....	....
Electrode no. 2	....	....	....
Electrode no. 3	....	....	....
Electrode no. 4	....	....	....
Electrode no. 5	....	....	....
Electrode no. 6	....	....	....
Electrode no. 7	....	....	....
Electrode no. 8	....	....	....
Electrode no. 9	....	....	....
Electrode no. 10	....	....	....

Resistance to earth of complete network ....

Resistances compared to previous readings   ☐ yes   ☐ no

Overall condition system   ☐ satisfactory   ☐ unsatisfactory

Repair work needed on ....

Inspecting engineer   Name ....   Signature ....   Date ....

Test results

	Location	Condition	Resistance to earth Ω
Electrode no. 1	....	....	....
Electrode no. 2	....	....	....
Electrode no. 3	....	....	....
Electrode no. 4	....	....	....
Electrode no. 5	....	....	....
Electrode no. 6	....	....	....
Electrode no. 7	....	....	....
Electrode no. 8	....	....	....
Electrode no. 9	....	....	....
Electrode no. 10	....	....	....

Resistance to earth of complete network ....(Ω)

Resistances compared to previous readings   ☐ yes   ☐ no

Remarks

▶ **Figure 11.6** *continued*

Environmental protection

(This information is correct at the time of going to press.)

12.1 The WEEE Directive and Regulations

The *Waste Electrical and Electronic Equipment (Producer Responsibility) Regulations 2004* ('the WEEE Regulations') implement provisions of the European Parliament and Council Directive on *Waste Electrical and Electronic Equipment* (2002/96/EC) ('the WEEE Directive').

The Directive aims to reduce the amount of such waste arising, to encourage reuse, recycling and recovery and to improve the environmental performance of all operators involved in the life cycle of electrical and electronic equipment. The Directive sets requirements relating to criteria for the collection, treatment, recycling and recovery of such waste. It makes producers responsible for financing most of these activities. Retailers/distributors also have responsibilities in terms of the take-back of waste and the provision of certain information. Private householders should to be able to return complete items of waste electrical and electronic equipment without charge.

The regulations apply to all electrical and electronic equipment placed on the market in the United Kingdom falling into any of ten product categories set out in Schedule 1, unless the equipment is part of another type of equipment which does not fall into any of these categories. The regulations also specify a voltage range into which the products in the ten categories must fall to be covered by the scope. This is up to 1000 V a.c. or up to 1500 V d.c.

▶ **Figure 12.1** Draft guidance from the DTI on the WEEE Directive

▶ **Table 12.1** Schedule 1 – Categories of electrical and electronic equipment covered by the regulations

Category	Examples of equipment
Large household appliances	large cooling appliances refrigerators, freezers other large appliances used for refrigeration, conservation and storage of food washing machines clothes dryers dish washing machines cooking, electric stoves, electric hot plates microwaves other large appliances used for cooking and other processing of food electric heating appliances, electric radiators other large appliances for heating rooms, beds, seating furniture electric fans air conditioner appliances other fanning, exhaust ventilation and conditioning equipment

continues

▶ **Table 12.1** *continued*

Category	Examples of equipment
Small household appliances	vacuum cleaners, carpet sweepers and other appliances for cleaning appliances used for sewing, knitting, weaving and other processing for textiles irons and other appliances for ironing, mangling and other care of clothing toasters and fryers grinders, coffee machines and equipment for opening or sealing containers or packages electric knives appliances for hair-cutting, hair drying, tooth brushing, shaving, massage and other body care appliances clocks, watches and equipment for the purpose of measuring, indicating or registering time scales
IT and telecommunications equipment	centralised data processing: mainframes minicomputers personal computers (CPU, mouse, screen and keyboard included) laptop computers (CPU, mouse, screen and keyboard included) notebook computers notepad computers printers, printing units copying equipment electrical and electronic typewriters pocket and desk calculators and other products and equipment for the collection, storage, processing, presentation or communication of information by electronic means user terminals and systems facsimile, telex, telephones, pay telephones, cordless telephones, cellular telephones, answering systems and other products or equipment transmitting sound, images or other information by telecommunications
Consumer equipment	radio sets television sets video cameras video recorders hi-fi recorders audio amplifiers musical instruments and other products or equipment for the purpose of recording or reproducing sound or images, including signals or other technologies for the distribution of sound and image by telecommunications
Lighting equipment	luminaires with fluorescent lamps with the exception of luminaires in households straight fluorescent lamps compact fluorescent lamps high intensity discharge lamps, including pressure sodium lamps and metal halide lamps low pressure sodium lamps other lighting or equipment for the purpose of spreading or controlling light with the exception of filament bulbs

Table 12.1 *continued*

Electrical and electronic tools (with the exception of large-scale stationary industrial tools)	drills saws sewing machines equipment for turning, milling, sanding, grinding, sawing, cutting, shearing, drilling, making holes, punching, folding, bending or similar processing of wood, metal and other materials tools for riveting, nailing or screwing or removing rivets, nails, screws or similar uses tools for welding, soldering or similar use equipment for spraying, spreading, dispersing or other treatment of liquid or gaseous substances by other means tools for mowing or other gardening activities
Toys, leisure and sports equipment	toys, leisure and sports equipment electric trains or car racing sets hand-held video game consoles video games computers for biking, diving, running, rowing etc. sports equipment with electric or electronic components coin slot machines
Medical devices (with the exception of all implanted and infected products)	radiotherapy equipment cardiology dialysis pulmonary ventilators nuclear medicine laboratory equipment for in-vitro diagnosis analysers freezers fertilisation tests other appliances for detecting, preventing, monitoring, treating, alleviating illness, injury or disability
Monitoring and control instruments	smoke detectors heating regulators thermostats measuring, weighing or adjusting appliances for household or as laboratory equipment other monitoring and control instruments used in industrial installations (e.g. in control panels)
Automatic dispensers	automatic dispensers for hot drinks automatic dispensers for hot or cold bottles or cans automatic dispensers for solid products automatic dispensers for money all appliances which deliver automatically all kind of products

The regulations do not apply to filament light bulbs, household luminaires, large-scale stationary industrial tools, implanted medical equipment and infected medical equipment at end-of-life and equipment intended specifically to protect the UK national interest and for a military purpose, e.g. arms, munitions and war material.

12.2 Responsibilities placed on the producer

A person who manufactures, resells or imports or exports electrical and electronic equipment, a producer, must be registered and will be required to meet all the following requirements.

i A producer will be required to report UK sales data in order for their market shares to be calculated.

ii A producer will have a responsibility for financing the collection, recovery and recycling of separately collected WEEE allocated to them according to their market shares. They must report evidence of its treatment at authorised treatment facilities, according to the agencies' treatment guidance. They must also report evidence that they have met the Directive's recovery and recycling/reuse targets for the separately collected WEEE allocated to them.

iii A producer will also be required to mark new equipment put onto the UK market, according to the Directive's requirements; and to provide certain information, as far as this is needed, on types of new equipment they put on the market, to facilitate the treatment and recovery of WEEE.

iv A producer supplying new equipment to business users will need to finance the treatment, recovery and disposal of the waste arising from this equipment unless alternative arrangements are made with business users. A producer who has supplied equipment to business users has this responsibility for this equipment if it is discarded when new replacement like for like equipment is supplied. If there is no replacement purchase, the business user is responsible for financing the treatment and recovery of the equipment purchased.

v Whichever party takes responsibility will need to report evidence of its collection, treatment and recovery according to the Directive's recovery and recycling/reuse targets.

A person who manufactures, resells, imports or exports electrical and electronic equipment including distance sellers must be registered with the National Clearing House (NCH) on behalf of the Secretary of State for Trade and Industry and a registration fee will be payable.

A producer may decide to belong to a compliance scheme and any such scheme must register all its members.

A producer, or their scheme, will be required to provide data to the NCH acting on behalf of the Government. The data will enable the NCH to allocate separately collected WEEE, according to each of the ten product categories. The Government is obliged to report on this basis to the European Commission. The data formats will require reporting of data on the numbers of electrical and electronic products and their weight put onto the market in the UK during any one year.

The Government will advise the formats for reporting on sales. It proposes that the total numbers and weights of products which businesses put onto the UK market should be reported to the NCH and used for its calculation of market shares in order to determine allocations of WEEE.

An importer will be obliged to report data on electrical and electronic equipment that is put onto the UK market and a distance seller will be obliged to report data on such equipment exported to other member states. The data will be provided on a confidential basis. It will be available to the NCH and the environment agencies but will not be made available to third parties. The Government is obliged to report aggregated data, by the product categories, to the European Commission.

A producer must provide at least for the financing of the collection, treatment, recovery and environmentally sound disposal of the proportion of WEEE from private households deposited at collection facilities and must furnish a certificate of compliance in respect of his recovery obligations. The Government is implementing the WEEE Directive's provisions on financing separately collected WEEE on the basis of what is essentially a market share approach, in which producers are allocated their 'WEEE shares'. Within this, there is flexibility for producers to discharge their obligations independently by arranging collection, treatment and recovery of WEEE from their own or others' products; or collectively; or by joining a compliance scheme.

12.3 Responsibilities placed on the retailer or distributor

Retailers and distributors of electrical and electronic equipment are obliged to:

▶ provide free in-store take-back of WEEE on sale of new like for like equipment, or to provide alternative arrangements to last holders of WEEE, via a compliance scheme approach; retailers and distributors providing in-store take-back need to ensure that the WEEE they collect is delivered to a designated collection facility to enable the WEEE to be sent for treatment and recovery
▶ ensure that private householders are informed of WEEE take-back facilities available to them and encouraged to participate in the separate collection of WEEE.

A distributor who supplies new equipment must ensure that waste electrical and electronic equipment can be returned to him on a one-to-one basis at least free of charge, provided that the electrical and electronic equipment that is returned is of equivalent type to, and has fulfilled the same functions as, the supplied equipment. The distributor can refuse to accept the returned waste electrical and electronic equipment if it presents a health and safety risk to personnel because of contamination.

12.4 Information provided by the producer

A producer is required to make arrangements to respond to requests for information to assist with the reuse, recycling and recovery of types of new equipment. Information for these products should be available within a year of being put on the UK market. The aim is to ensure that information is provided to facilitate the reuse, recycling and recovery of the equipment. It may be reasonable to be asked to provide advice on the location of items and substances covered by the Environment Agency's guidance on the treatment of WEEE and the location and type of dangerous/hazardous substances, materials or components within the type of equipment.

There is flexibility for producers to decide how best to make available information on these kinds of issues. They may wish to respond to requests for information, when these arise, or to take steps to make information available. Electronic means of information dissemination such as websites or CD-ROMs may provide the most effective and least-cost compliance. Products may carry direct information marking.

▶ **Figure 12.2** Symbol for electrical items that should not be codisposed with other waste

12.5 Marking

The WEEE regulations require producers to ensure that equipment which they put on the UK market is marked with the crossed out wheeled bin symbol shown in Figure 12.2. The symbol must be printed visibly, legibly and indelibly.

The standards body, CENELEC, is working on a standard for the size and durability of this symbol. The Government will advise the reference number of this standard once it is agreed and published in the *Official Journal of the European Communities*.

12.6 Take-back for consumers

The regulations require all retailers and distributors of electrical and electronic products to domestic consumers to provide free take-back in store to enable consumers to return their WEEE, when making a like for like purchase of new equipment. This means that consumers will have the right to take back their old product free of charge when going to a shop to buy a new like for like product. 'Like for like' is understood to mean equipment that is of equivalent type or fulfils the same function. This means a customer might expect to be able to take back an old personal compact disc player when buying a new one. It also implies that an old cassette player could be returned when a compact disc player is being purchased, because both are used for the function of playing recorded music and sound. The take-back obligations apply to all retailers, both those who specialise in selling electrical and electronic equipment and other retailers who also sell some electrical or electronic equipment as well as other goods.

The regulations provide retailers with a choice of compliance route for discharging this take-back obligation. Retailers and other distributors of electrical and electronic equipment are expected to offer 'direct' take-back services as outlined above or, alternatively, 'indirectly' to show participation in a compliance scheme which would offer WEEE take-back services to consumers. A retailer compliance scheme or schemes would be subject to approval by the Government.

12.7 Take-back for retailers and distributors

Retailers and distributors are also entitled to free take-back of WEEE they collect in store if they decide to follow this route. If they are providing direct take-back services, for example, in store, they should deposit the items they have collected at a designated collection facility. Either the retailer scheme or the NCH will signpost retailers and distributors holding items collected via their own direct, in-store take-back services to the local nearest designated collection facility.

12.8 In-store take-back

Where retailers choose to comply with their take-back obligation in store, this service should be offered for all of the electrical and electronic equipment sold from a particular retail premises. It is recognised that large or bulky goods, like refrigerators or television sets, although marketed via retail premises, are often delivered to customers and an old product is collected when that delivery is made. Many retailers currently offer this arrangement as a service to their customers and this is taken into account in the

Regulations. The Government appreciates that there will be continuing consumer demand for this service and that retailers will want to respond to this. These collection on delivery arrangements may be used in lieu of in-store take-back provided the collection is free.

Distance sellers based in the UK, such as internet sellers, who do not sell through retail premises, will be expected to offer an alternative free take-back service in lieu of in-store take-back, or to join a retailer/distributor compliance scheme.

Some discarded electrical and electronic equipment will be classified as hazardous waste. DEFRA will soon be issuing, for consultation, new regulations setting out procedures for movements of hazardous waste in England and Wales. These will be available on DEFRA's website.

12.9 Retailer/distributor compliance scheme

The regulations provide for retailers and distributors to comply with the take-back obligation by showing they are participating in an approved retailer compliance scheme. A compliance scheme must be approved by the Secretary of State for Trade and Industry prior to registration. A proposal for approval of a scheme may be submitted at any time after the Regulations have entered into force.

The primary objective of a scheme should be to support an adequate UK-wide WEEE collection network – as an alternative to its members offering direct, in-store take-back services.

Retailer-led collection services may take the form of take-back points in major retail parks, or there may be other arrangements better suited to particular areas. It will be for the scheme to define optimum arrangements, consistent with achieving UK-wide coverage. Where facilities in major shopping centres are established, they must obtain the relevant waste management licence/permit or exemption and must be in accordance with local planning consents and tenancy agreements. A retailer/distributor compliance scheme would be expected to:

▷ provide and resource WEEE take-back services to meet its members' take-back obligations under the regulations
▷ provide an adequate UK network as an alternative to its members offering direct, in-store take-back, offering an equivalent level of availability compared to in-store take-back to last holders
▷ provide funding support for the upgrade of WEEE collection at civic amenity sites, in particular, but not exclusively to assist them to meet the minimum standards for designated collection facility status
▷ provide plans for complementary WEEE collection services and facilities to be managed by the scheme itself, ensuring UK-wide coverage for the network
▷ maintain records of its membership and report this annually to the environment agencies
▷ take steps to ensure proportionate treatment of small retailers who want to join the scheme, particularly those whose principal business is not selling electrical and electronic equipment
▷ make available to householders information on WEEE collection facilities – see Section 12.10 of this guidance on information provision.

A retailer/distributor compliance scheme will need to register with the environment agencies for the purposes of enforcement. A registration fee will be payable. A retailer/distributor compliance scheme will be subject to approval by the Secretary of State.

A proposal seeking approval of a retailer/distributor compliance scheme should ensure that the scheme meets the following criteria:

i The scheme should set out an operational plan for a network of WEEE takeback services and facilities, which it will offer as an alternative to its members offering direct, in-store take-back. This network should offer an equivalent level of access and availability compared to instore take-back to last holders.

ii The scheme should indicate how it will arrange appropriate funding support for upgrades to the civic amenity site network. Collection facilities within the scheme would be expected to meet certain minimum standards.

iii The scheme should provide plans for complementary WEEE take-back services and facilities to be managed by the scheme itself, taking account of the considerations indicated above: ensuring UK-wide coverage.

iv The scheme should provide information on its management and membership fee structures.

v The scheme should provide a plan for how it will make available information on WEEE take-back facilities to householders.

12.10 Provision of information to consumers

The regulations require retailers and distributors offering take-back services to ensure that private householders are informed of the WEEE take-back facilities available to them. Householders should be encouraged to participate in the separate collection of WEEE. This information may cover retailers' own collection services or facilities and any other systems available for take-back of WEEE. Households should be informed of the meaning of the crossed out wheeled bin symbol on products covered by the directive (i.e. that WEEE should not be codisposed with other waste). It may be appropriate for retailers to liaise with local authorities, charities, the waste management industry and others in discharging this obligation to make information available. It will be for the parties concerned, either retailers complying individually or a compliance scheme, to determine the best ways to communicate this information to householders. The options for this may include:

▶ posters and/or leaflets in stores/point of sale
▶ notices at civic amenity and other collection sites
▶ information on websites
▶ advertisements on delivery vehicles or refuse collection vehicles
▶ advertisements in the local media
▶ collaborative information exercises with local authorities.

12.11 Collection and storage of WEEE

Sites offering facilities for separately collecting WEEE may register with the NCH as 'designated' collection facilities for the purposes of securing free collection by producers of separately collected WEEE. These sites may be, for example:

▶ local authority civic amenity sites or transfer stations
▶ retailer returns/bulking points associated with collection on delivery
or
▶ collection/bulking points associated with a retailer compliance scheme.

In some instances, designated collection facilities may be nominated as points for collections by producers, but not offer collection facilities for householders. An example might be a waste transfer station operated as part of the local civic amenity infrastructure. Collection facilities will be invited to register with the NCH details of their operators, locations, access arrangements and stipulate whether they offer facilities for all or some of the WEEE collection 'groupings'. These groupings are proposed as:

▸ WEEE Directive Annex 1A, categories 1 and 10 – large domestic appliances (this includes refrigerators, freezers, which are, in practice, collected and stored at collection sites separately already to facilitate adherence to the ozone depleting substances regulations)
▸ Annex 1A, categories 3 and 4 – IT/telecoms, consumer equipment
▸ Annex 1A categories 2,6,7,9 – small household appliances, electrical and electronic tools, toys, leisure and sports equipment, and monitoring and control equipment
▸ Annex 1A category 5 – lighting equipment.

The WEEE Directive's annex III includes some technical requirements covering sites used for the storage of WEEE prior to treatment. These requirements are for impermeable surfaces (with provision for spillage collection facilities and, where appropriate, decanters and cleanser/degreasers) and weather-proofing for appropriate areas. Sites will be expected to meet these standards. The regulations provide for designated collection facilities to meet these standards, as well as to have appropriate containers and suitable signage to guide users. Sites applying for designated collection facility status should review their facilities in the light of these requirements. The Government proposes that the National Clearing House should undertake registration of designated collection facilities for participation in its free collection service. The NCH should write to local authorities and waste management authorities as early as possible to invite this.

The Government understands that the NCH allocation mechanism should be able to assimilate new designated collection facilities into the system.

12.12 Rural areas

Rural areas of the UK, such as the Highlands of Scotland, central Wales, western areas of Northern Ireland or the south west of England, present their own challenges in terms of the separate collection of WEEE. However, there should be operational solutions. It may be appropriate for local waste management authorities to apply to the NCH to designate several WEEE collection facilities together and to ask for a collection schedule which includes a 'round robin' pattern of pick-ups – as is commonly used for other services to such areas. The Government also expects retailers and distributors of electrical and electronic equipment to play their part in meeting these challenges to provide adequate take-back facilities across the whole of the UK.

12.13 Permitting of WEEE

WEEE treatment activities must be dealt with in accordance with the permitting system which is operated under the *Waste Framework Directive* (75/442/EEC). Any establishment or undertaking carrying out treatment operations must obtain a permit or licence from the competent authority. However, the WEEE Directive imposes some additional requirements under Article 6 in respect of facilities that carry out treatment of WEEE which must be transposed to UK legislation. Treatment shall as a minimum include the removal of all fluids and selective treatment in accordance with Annex II of the Directive. Annex II lists various preparations and components that have to be removed and treated in such a way that environmentally sound reuse and recycling of components or whole appliances is not hindered. Treatment should also use best available treatment, recovery and recycling techniques. Storage and treatment sites at treatment facilities have to meet the technical requirements set out in Annex III of the WEEE Directive. The permitting requirements relating to the treatment of Waste Electrical and Electronic Equipment (WEEE) under Article 6 of Directive 2002/96/EC are expected to be transposed into law by the *Environmental Permitting Regulations*, being produced by DEFRA, which will replace the majority of Part II of the *Environmental Protection Act 1990* and the *Waste Management Licensing Regulations 1994*.

12.14 Ecodesign

The WEEE Directive requires the Government to encourage the ecodesign of electrical and electronic equipment in order to make it easier to reuse, recycle and recover WEEE,

▶ **Figure 12.3** A decision tree that could be used to decide whether or not a product might come within the scope of the WEEE regulations (annex B)

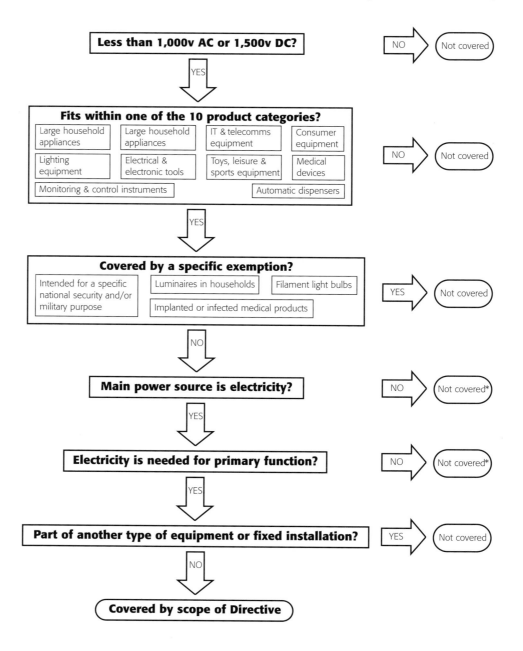

* While these exclusions are not expressly provided for in the Directive, it is the UK view that they apply. Producers should rely on independent legal advice on compliance.

its components and materials. It is also required to prevent the use of design techniques or manufacturing processes that prevent the reuse or recycling of WEEE, unless there are over-riding environmental or safety advantages.

12.15 Further information

Enquiries regarding the WEEE regulations should be referred to the Environment Agency in England and Wales, SEPA in Scotland and the Environment and Heritage Service in Northern Ireland. Contact details are as follows:

Environment Agency
Rio House, Waterside Drive
Aztec West, Almondsbury
Bristol
BS32 4UD
tel: 08459 333111
email: enquiries@environment-agency.gov.uk
www.environment-agency.gov.uk

Scottish Environment Protection Agency (SEPA)
Erskine Court
The Castle Business Park
Stirling
FK9 4TR
tel: 01786 457700
fax: 01786 446885
www.sepa.org.uk

Producer Responsibility Unit
Environment and Heritage Service
Commonwealth House
35 Castle Street
Belfast
BT1 1GU
tel: 028 9054 6484
fax: 028 9054 6480
email: weee@doeni.gov.uk
www.ehsni.gov.uk

Guidance documents are available from these addresses or from the Regional Environment Agency Office.

Environmental helpline:

0800 585 794

The helpline is a Government telephone enquiry service providing a comprehensive information and signposting service for firms seeking advice on a wide range of environmental issues that may affect their business. Case studies and guides to help with various issues are available.

The WEEE Regulations 2004 S.I.[ ] are available from the Stationery Office and online:

www.hmso.gov.uk

Draft guidance notes are available from:

DTI Publications Orderline
Admail 528
London
SW1W 8YT
tel: 0870 1502 500
fax: 0870 1502 333
www.dti.gov.uk

Comments on guidance notes should be addressed to DTI's Sustainable Development Unit:

Department of Trade and Industry
Sustainable Development Unit
151 Buckingham Palace Road
London
SW1W 9SS
tel: 020 7215 1036
fax: 020 7215 5835
www.dti.gov.uk/sustainability/weee/index.htm

Department for Environment, Food and Rural Affairs
Producer Responsibility Unit
Zone 7/F9 Ashdown House
123 Victoria Street
London SW1E 6DE
tel: 020 7082 8778
fax: 020 7082 8764
www.defra.gov.uk/environment/waste/topics/electrical

Legionellosis 13

To comply with their legal duties, employers and those with responsibilities for the control of premises should:

a Identify and assess sources of risk – this includes checking whether conditions are present which will encourage the bacteria to multiply, which are:
- is there water at a temperature between 20 and 45 °C?
- is there a means of creating and disseminating breathable droplets such as the aerosol created by a shower or a cooling tower?
- are there susceptible people who may breathe in the contaminated aerosols?

b Prepare a scheme for controlling the risks.
c Implement, manage and monitor precautions.
d Keep records of the precautions, and
e Appoint a person to be responsible.

Figure 13.1 Legionellosis legal duties

13.1 Introduction

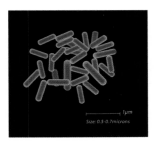

Figure 13.2 The bacteria *Legionella pneumophila*

Legionnaires' disease is a potentially fatal form of pneumonia. It was named after an outbreak of severe pneumonia which affected a meeting of the American Legion in 1976. It is an uncommon but serious disease, which occurs more frequently in men than women. It usually affects middle-aged or elderly people and it more commonly affects smokers or people with other chest problems. Legionnaires' disease is uncommon in younger people and is very uncommon under the age of 20. About half the cases of legionnaires' disease are caught abroad. The other half are the result of infections acquired in the UK.

The germ which causes legionnaires' disease is a bacterium called *Legionella pneumophila*. People catch legionnaires' disease by inhaling small droplets of water suspended in the air which contain the *Legionella* bacterium. However, most people who are exposed to *Legionella* do not become ill. Legionnaires' disease does not spread from person to person.

On average there are 200–250 cases of legionnaires' disease reported annually in the UK.

The bacterium that causes legionnaires' disease is widespread in nature. It mainly lives in water, for example ponds, where it does not usually cause problems. Outbreaks occur from purpose-built water systems where temperatures are warm enough to encourage growth of the bacteria, e.g. in cooling towers, evaporative condensers and whirlpool spas, and from water used for domestic purposes in buildings such as hotels.

Most community outbreaks in the UK have been linked to installations such as cooling towers or evaporative condensers which can spread droplets of water over a wide area. These are found as part of air-conditioning and industrial cooling systems.

To prevent the occurrence of legionnaires' disease, companies which operate these systems must comply with regulations requiring them to manage, maintain and treat them properly. Among other things, this means that the water must be treated and the system cleaned regularly.

The symptoms of legionnaires' disease are similar to those of the flu:

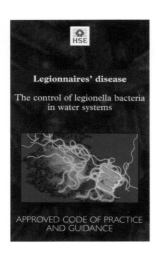

- high temperature, feverishness and chills
- cough
- muscle pains
- headache; leading on to
- pneumonia (very occasionally)
- diarrhoea and signs of mental confusion.

The illness is treated with an antibiotic called erythromycin or a similar antibiotic.

The guidance given in this chapter is based on Health and Safety Commission Publication: *The prevention or control of Legionellosis (including Legionnaires' Disease) reference L8(REV)*. Detailed technical advice on assessing and minimising the risk from exposure to *Legionellae* is given in Health and Safety Executive Guidance HS(G)70 *The control of Legionellosis including Legionnaires' Disease*.

▶ **Figure 13.3**
Legionnaires' disease.
Approved Code of
Practice. L8

13.2 Legislation

The *Management of Health and Safety at Work Regulations 1999* (Regulation 6), the *Control of Substances Hazardous to Health Regulations 1994* (Regulation 6) and the *Health and Safety at Work etc. Act 1974* (Sections 2, 3 and 4) have requirements for preventing or minimising the risk from exposure to *Legionellae*. These include requirements for:

- identification and assessment of risk
- preventing or minimising the risk from exposure
- management, selection and training of personnel
- the need for record keeping, and
- responsibilities on designers, manufacturers, importers, suppliers and installers.

The *Reporting of Injuries, Diseases and Dangerous Occurrences Regulations 1995* (RIDDOR) require employers and others, e.g. the person who has control of the premises, to report to HSE accidents and diseases that arise out of or in connection with work. Cases of legionellosis are reportable under RIDDOR if a doctor notifies the employer and if the employee's current job involves working on or near cooling systems that use water or hot water service systems in the workplace.

Those who have, to any extent, control of premises, have a duty under the *Notification of Cooling Towers and Evaporative Condenser Regulations 1992* to notify the local authority in writing with details of notifiable devices. These consist of cooling towers and evaporative condensers, except where they contain water that is not exposed to the air and the water and electricity supply are not connected. If a tower becomes redundant and is decommissioned or dismantled, this should also be notified. The main purpose of these regulations is to help in the investigation of outbreaks.

13.3 Water systems creating risk

The legislation applies to any work activity or premises connected with a trade, business or undertaking where water is used or stored and where there are means of creating water droplets which may be inhaled, thereby causing a reasonably foreseeable risk of legionellosis. Effective precautions to stop or contain transmission of droplets over an area where they might be inhaled would prevent risk. Experience has shown that the following present a risk of legionellosis:

▸ water systems incorporating a cooling tower
▸ water systems incorporating an evaporative condenser
▸ hot water services
▸ the hot and cold water services, irrespective of size, in premises where occupants are particularly susceptible, e.g. old people's homes, nursing homes and general health care premises
▸ humidifiers and air washes which create a spray of water droplets and in which the water temperature is likely to exceed 20 °C and contain a source of nutrients (sludge, scale, rust, algae and other organic matter)
▸ spa baths and whirlpools in which warm water is deliberately agitated and recirculated
▸ general plant and systems containing water likely to exceed 20 °C and which may release a spray or aerosol during operation or when being maintained
▸ shower heads.

If any employer or person responsible for a trade, business, or similar undertaking has water systems as described above, arrangements must be made for an assessment to be carried out to identify the risk to health and the measures that can be taken to prevent or control the risk of exposure to *Legionellae*.

13.4 Identification and assessment of risk

For premises with systems as described above, a suitable and sufficient assessment should be carried out to identify and assess the risk of legionellosis from work activities and water sources. The purpose of the assessment is to enable valid decisions to be made with respect to the risks to health and what measures for prevention or control should be taken.

13.5 Preventing or minimising the risk

Where a reasonably foreseeable risk has been identified, the use of water systems, etc. that could lead to exposure should be avoided, so far as is reasonably practicable, until measures have been taken to minimise the risk. The scheme to minimise the risk should be specific and sufficiently detailed to enable it to be implemented and managed. It must contain information concerning the plant or system necessary to minimise the risk from exposure and ideally should include:

▸ an up-to-date plan, showing the layout of the plant or system
▸ a description of the correct and safe operation of the plant or system
▸ the precautions to be taken.

The precautions to minimise the risk would include the following:

▶ minimisation of the release of water spray
▶ avoidance of water temperatures and conditions that favour the growth of *Legionellae* and other microorganisms
▶ avoidance of water stagnation
▶ avoidance of the use of materials that harbour bacteria and other microorganisms or provide nutrients for microbial growth
▶ maintenance of a clean system and the water in it
▶ use of water treatment techniques
▶ action to ensure correct and safe operation and maintenance.

Any scheme must include control measures and the necessary checks to ensure that the control is effective.

13.6 Control

Where assessment has shown there is a reasonable foreseeable risk, the employer, self-employed person or person in control of the premises should appoint a person to take managerial responsibility for the implementation of precautions.

13.7 Record keeping

The responsible person must ensure that suitable records are kept, including:

▶ the name and position of the person having managerial responsibility
▶ the assessment of the risk, and the name and position of the person who carried out the assessment
▶ a written scheme for minimising or eliminating the risk
▶ the names and positions of the persons responsible for implementing the scheme
▶ method of management of the scheme
▶ the plant systems that are in use and their state, for example if they are drained down.

Records should be kept with respect to the implementation of the scheme. The records should show:

▶ the precautionary measures taken, with clear indication that they have been correctly carried out, when and by whom
▶ the results of any inspection, test or check carried out, including the dates and the names of the persons carrying out the checks
▶ a register of remedial works necessary, including the date the work was completed.

13.8 Practical implementation of the control of *Legionella* in water systems

13.8.1 Cooling towers and evaporative condensers

Evaporative cooling systems dissipate heat using water as the heat exchange medium. Such systems provide a suitable environment for the growth of many microorganisms including *Legionella* which can be spread widely by aerosol into the area around the cooling tower.

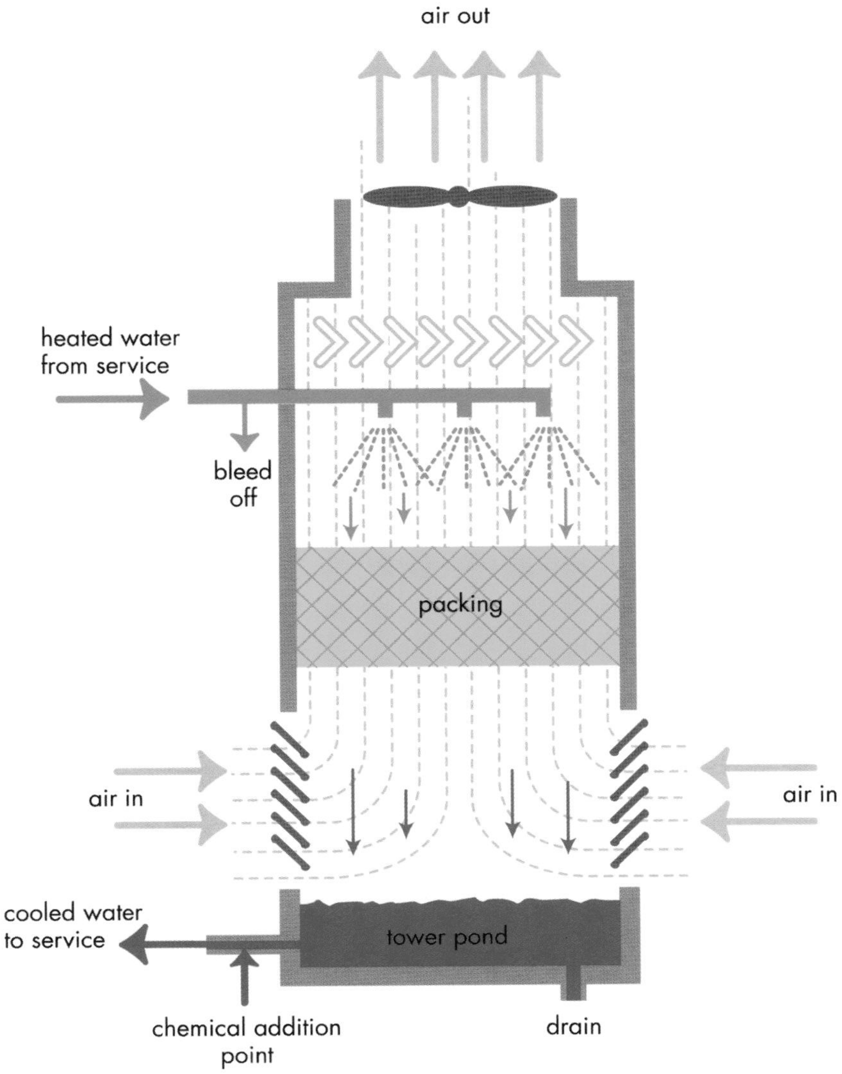

Figure 13.4 Induced
draft cooling tower

In some circumstances it may be possible to use alternative methods of cooling, such as air blast coolers which will avoid the risks presented by a wet cooling tower or evaporative condenser.

Cooling tower or evaporative condenser	Dry cooling system
water treatment chemicals required	can be larger
regular cleaning required	heavier
regular disinfection required	noisier
monitoring required	less economic in some cases
management system required	less maintenance required

Table 13.1
Comparison table

Design

Dry cooling systems should be considered for new systems and when cooling towers are due to be replaced.

Wet systems must be designed so that drift is minimised and that operation, cleaning and disinfection can be performed safely. (Drift is circulating water lost from the tower as liquid droplets suspended in the exhaust air stream.) Drift eliminators, usually made of plastic or metal, should be installed in all towers and, although they do not 'eliminate' drift, they can significantly reduce it. Older wooden slats should be replaced.

The volume above the cooling tower pond should be well enclosed to reduce the effects of windage and it may also be necessary to screen the tower or pond to prevent the entry of birds, vermin, leaves, debris and reduce solar gain.

The water distribution system must be designed to create as little aerosol as possible.

The water system must be designed to avoid deadlegs and 'difficult-to-drain' loops and bends. Undisturbed water means microbes can grow and multiply and then rapidly colonise the water system. Schematics of the water system should be available.

Those parts of the tower that get wet should be accessible for cleaning and should be shaded to avoid growth of algae.

The pond should have a sloping floor with a large drain connection able to carry away water and slurry easily. Supplementary drain valves may be needed for individual items of equipment.

Towers should be constructed of materials that can be easily cleaned and disinfected.

Make-up water that is not mains supplied may need pre-treatment.

A full water treatment programme must be integrated into the system design, with provision for sample, dosing, injection bleed and drain points. Dosing and bleeding should, ideally, be automated.

Cooling towers should be positioned as far away as possible from air-conditioning and ventilation inlets, opening windows and occupied areas taking account of prevailing wind directions. Old towers may be able to be relocated to more suitable areas when being replaced.

Management

The management control system must include all the elements of the cooling system including the cooling tower, evaporative condenser, recirculating pipework, heat exchanger, pumps, supply tanks and pre-treatment equipment.

During commissioning or recommissioning, the system should not be filled and left inoperative for periods exceeding a week. It should be drained down and refilled. The water should include a biocide. Results of commissioning should be included in the operating and maintenance manual.

Operation

Cooling systems and towers should be kept in regular use. If this is not possible, the system should be run once a week and dosed with water treatment chemicals, and the water quality monitored. The system should be run long enough to distribute the treated water. If the system is out of use for a period between one week and one month, the water should be treated immediately with a biocide on reuse. If the system is out of use for longer periods, refer to the Approved Code of Practice for details.

Operating manuals should be available. They should be easily understandable and include system drawings, details, water treatment programmes and records.

Operating manuals should include a detailed maintenance schedule.

Treatment programmes

A complete water treatment programme based on the physical and operating parameters for the cooling system and a thorough analysis of the make-up water should be established. The components of the water treatment programme should be environmentally acceptable and comply with any local discharge requirements. Any treatment programme must take account of corrosion, scale, fouling and microbiological activity. All components of the treatment programme should preferably be dosed by pump or ejector, thus minimising health and safety risks to operators and ensuring rates of application are maintained.

Corrosion of steel can lead to conditions which encourage the growth of *Legionella*. Scale formation reduces efficiency and can protect the bacteria from the effectiveness of the biocidal treatment. Fouling refers to the deposition of particulate material and debris such as dirt, dust, leaves, pollen and insects leading to increased microbiological activity and the proliferation of *Legionella*. Cooling systems provide an environment in which microbiological systems, including *Legionella*, can proliferate. Water temperatures, pH conditions, presence of oxygen, carbon dioxide, sunlight, nutrients and large surface areas favour the growth of microorganisms.

Monitoring programmes

The cooling and make-up water should be routinely monitored to ensure continued effectiveness of the treatment programme, the minimum frequency being once a week.

▶ **Table 13.2** Typical on-site monitoring checks

Parameter	make-up water checked:	cooling water checked:
Calcium hardness as mg/l $CaCO_3$	monthly	monthly
Magnesium hardness as mg/l $CaCO_3$	monthly	monthly
Total hardness as mg/l $CaCO_3$	monthly	monthly
Total alkalinity as mg/l $CaCO_3$	quarterly	quarterly
Chloride as mg/l Cl	monthly	monthly
Sulphate as mg/l SO_4	quarterly	quarterly
Conductivity μs (total dissolved solids)	monthly	weekly
Suspended solids mg/l	quarterly	quarterly
Inhibitor(s) level mg/l	—	monthly
Oxidising biocide mg/l	—	—
Temperature °C	—	weekly
pH	quarterly	weekly
Soluble iron as mg/l Fe	quarterly	quarterly
Total iron as mg/l Fe	quarterly	quarterly
Concentration factor	—	monthly
Microbiological activity	quarterly	weekly
Legionella		**quarterly**

Microbiological activity is generally measured on a weekly basis using dip slides. A dip slide is a plastic slide coated with nutrient sugar. A slide is dipped into the water, incubated usually for 48 h at a temperature of 30 °C which provides suitable conditions for bacteria to grow and then compared to a chart, thereby indicating the number of bacteria in the water. If the water treatment programme is effective, the dip slide counts should be consistently low.

▶ **Table 13.3** Action levels following microbial monitoring for cooling towers

Aerobic count cfu/ml at 30 °C (minimum 48 h incubation)	*Legionella* bacteria cfu/litre	Action required
10 000 or less	100 or less	system under control
more than 10 000 and up to 100 000	more than 100 and up to 1000	review programme operation; review control measures and risk assessment; reconfirm count by resampling
more than 100 000	more than 1000	implement corrective action; system should be resampled; system should be shot-dosed with a biocide. Review control measures and risk assessment to identify remedial actions

Monitoring for Legionella

The routine monitoring scheme should include periodic sampling for *Legionella*. More frequent sampling should be carried out during commissioning a system and when establishing a treatment programme, when sampling should be carried out on a monthly basis. If a *Legionella* positive sample is found, more frequent sampling should be instigated until it can be demonstrated the system is under control. The sampling should be in accordance with ISO 11731: 1998, should be performed as near to the heat source as possible and the biocide neutralised where possible. The sample should be tested by a United Kingdom Accreditation Service (UKAS)-accredited laboratory.

Legionella bacteria are commonly found in almost all natural water, so sampling may often yield positive results. Results need careful interpretation by an experienced microbiologist.

Cleaning and disinfection

Disinfection, cleaning and manual desludging of cooling towers should be undertaken:

▶ immediately before the system is first commissioned
▶ twice yearly and more often if the local environment is dirty
▶ after any prolonged shutdown of a month or longer (note that a risk assessment may indicate the need for cleaning and disinfection after a period of less than one month)
▶ if the tower or any part of the system has been altered
▶ if the cleanliness of the tower or system is in any doubt
▶ if microbiological monitoring indicates there is a problem.

Details on cleaning and disinfection are given in the ACOP and may require precautions such as performing the cleaning when the building is unoccupied and no members of the public are in the vicinity, ensuring all the windows are closed, the area tented and personnel performing the cleaning are adequately trained and wear suitable positive pressure respirators with full face piece.

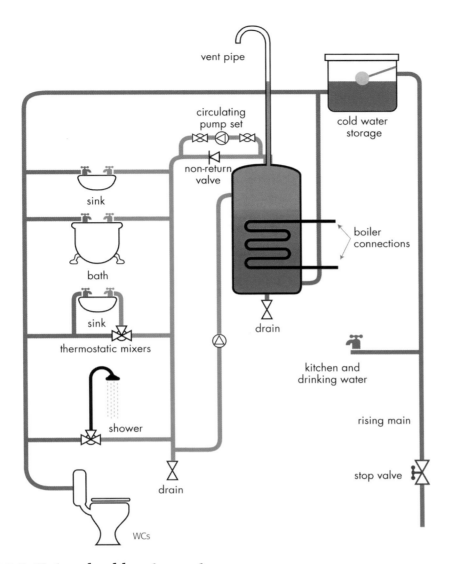

vent pipe

circulating
pump set

cold water
storage

non-return
valve

sink

boiler
connections

bath

sink

drain

thermostatic mixers

kitchen and
drinking water

shower

rising main

drain

stop valve

WCs

13.8.2 Hot and cold water systems

A wide variety of systems are available to supply hot and cold water systems and, in the past, have been associated with more reported outbreaks of legionnaires' disease than cooling towers. However, in recent years there have been very few outbreaks – probably due to better maintenance and care. *Legionella* will only grow in cold water systems and the distribution pipework when there are increased temperatures (for example due to heat gain), appropriate nutrients and stagnation.

Detailed information concerning the design and construction, management, treatment, monitoring and cleaning of hot and cold water systems is given in the HSE's publication: *Legionnaires' disease: the control of legionella bacteria in water systems.*

Design

The following points should be considered when designing hot and cold water systems:

▶ materials such as natural rubber, hemp, linseed-oil-based jointing compounds and fibre washers should not be used
▶ low corrosion materials such as copper, plastic, stainless steel should be used where possible
▶ water storage tanks should be fitted with covers, and insect screens fitted to pipework open to the atmosphere such as overflow and vent pipes

- multiple linked storage tanks should be avoided because of possible unequal flows and stagnation
- accumulator vessels on pressure boosted systems should be fitted with diaphragms which are accessible for cleaning
- point of use hot water heaters with no storage should be considered for remote or low use outlets
- showers, excluding safety showers, should not be installed if they are likely to be used less than once a week.
- thermostatic mixing valves should be sited as close to the point of use as possible
- thermostatic mixing valves should not be used with low volume spray taps in buildings with susceptible populations
- the hot water cylinder should be suitably sized to meet the daily use of the system without drop in hot water temperature
- the hot water vent pipe should not discharge hot water but should discharge into a tundish
- hot water cylinders should provide water at a temperature of at least 60 °C
- hot water distribution pipes should be insulated
- hot water pipework should reach a temperature of 50 °C within one minute from demand being placed
- low use cold water outlets should be installed upstream of high use outlets
- stored cold water should not exceed one day's use
- multiply connected tanks should be so interconnected as to avoid water stagnation
- cold water tanks should be kept cold by siting and insulation and kept away from sources of heat to prevent excessive temperature rises in the cold water, typically not more than 2 °C.

The risk from *Legionella* in hot and cold water systems must be managed. Procedures given in the Approved Code of Practice include:

- procedures to be followed upon commissioning or recommissioning a system
- procedures to be followed when operating a system
- maintenance procedures
- the need for regular flushing of showers and taps
- treatment and control programmes
- monitoring programmes
- procedures to be followed for cleaning and disinfecting a system.

Service	Task	Frequency
Hot water services	arrange for samples to be taken from hot water calorifiers in order to note condition of drain water	annually
	check temperatures in flow and return lines at calorifiers	monthly
	check hot water temperature for up to one minute to see if it has reached 50 °C in the sentinel taps	monthly
	visual check on internal surfaces of calorifiers for scale and sludge; check taps for temperature as above on a rotation basis	annually
Cold water services	check tank water temperature remote from ball valve and mains temperature at ball valve. Note maximum temperature by fixed maximum/minimum thermometers where fitted	six-monthly
	check that temperature is below 20 °C after running the water for up to two minutes in the sentinel taps	monthly
	visually inspect cold water storage tanks and carry out remedial work where necessary. Check representative taps for temperature as above on a rotation basis	annually
Shower heads	dismantle, clean and descale shower heads and hoses	quarterly or as necessary
Little-used outlets	Flush through and purge to drain, or purge to drain immediately before use without release of aerosols	weekly

▶ **Table 13.4**
Recommended inspection frequencies for hot and cold water services

13.8.3 Other systems that may present a risk of *Legionella*

Other systems producing aerosols that can present a risk of *Legionella* include spas, whirlpool baths, humidifiers and air washers. A spa is a bath or small pool where warm water is constantly recirculated, often through high velocity jets or with the injection of air to agitate the water. The water is not changed after each user, instead it is filtered and chemically treated. Spa baths can be a risk even when not being used by bathers, for example when being run for display purposes. Whirlpool baths do not present the same risk because the water is discharged after each use.

Atomising humidifiers and spray-type air washers may use water from tanks where the temperature exceeds 20 °C and unless they are regularly cleaned and maintained can become heavily contaminated. Note that 'portable' or 'room' humidifiers of the type that have a water supply that is sprayed or atomised into the room are not recommended in NHS premises.

▶ **Table 13.5**
Recommended
inspection frequencies
for other systems

Service	Task	Frequency
Ultrasonic humidifiers/foggers and water misting systems	if equipment is fitted with UV lights, check to ensure effectiveness of lamp (check to see if within working life) and clean filters	six-monthly or according to manufacturer's instructions
	ensure automatic purge of residual water is functioning	as part of machinery shutdown
	clean and disinfect all wetted parts	as indicated by risk assessment
	sampling for *Legionella*	as indicated by risk assessment
Spray humidifiers, air washers and wet scrubbers	clean and disinfect spray humidifiers/air washers and make-up tanks including all wetted surfaces, descaling as necessary	six-monthly
	confirm the operation of non-chemical water treatment (if present)	weekly
Water softeners	clean and disinfect resin and brine tanks. Check with manufacturers what chemicals can be used to disinfect resin bed	as recommended by manufacturer
Emergency showers and eye wash sprays	flush through and purge to drain	six-monthly or more frequently if recommended by manufacturer
Sprinkler and hose reel systems	when witnessing tests of sprinkler blow-down and hose reels ensure there is minimum risk of exposure to aerosols	as directed
Lathe and machine tool coolant systems	clean and disinfect storage and distribution system	six-monthly
Spa baths	check filters – sand filters should be back-washed daily	daily
	check water treatment – pools should be continuously treated with an oxidising biocide	three times daily
	clean and disinfect entire system	weekly
Horticultural misting systems	clean and disinfect distribution pipework, spray heads and make-up tanks including all wetted surfaces – descale as necessary	annually
Dental equipment	drain down and clean	at the end of each working day
Car and bus washes	check filtration and treatment system, clean and disinfect	see manufacturer's instructions
Indoor fountains and water features	clean and disinfect ponds, spray heads and make-up tanks including all wetted surfaces, descaling as necessary	interval depending on condition

13.9 References

Approved Code of Practice – The prevention or control of Legionellosis (including Legionnaires' Disease) – HSC L8(rev).

The control of legionellosis, including Legionnaires' Disease HS(G)70 (formerly guidance note EH48) – HSE, and the 1998 supplement *The control of legionellosis in hot and cold water systems* MISC150.

Legionellosis – An interpretation of the requirements of the Health and Safety Commission's Approved Code of Practice : The prevention and control of Legionellosis CIBSE GN3 : 1993 ISO 11731: 1998

Typical safety instructions

A.1 General provisions

A1.1 Scope
These safety instructions are for general application for work involving either, or both, non-electrical and electrical work as further described in Sections A.2 and A.3.

A.1.2 Definitions
For use with this Safety Instruction.

Appliance: a device requiring a supply of electricity to make it work.

Approved: sanctioned in writing by the responsible director in order to satisfy in a specified manner the requirements of any or all of the applicable safety rules.

Competent person: person required to work on electrical equipment, installations and appliances and recognised by the employer as having sufficient technical knowledge and/or experience to enable him/her to carry out the specified work properly without danger to themselves or others. It is recommended that this competence should be recognised by means of written documentation.

Conductor: an electrical conductor arranged to be electrically connected to a system.

Customer: a person, or body, that has a contractual relationship with the employer for the provision of goods or services.

Danger: risk of injury to persons (and livestock where expected to be present) from:

 i fire, electric shock and burns arising from the use of electrical energy, and
 ii mechanical movement of electrically controlled equipment, insofar as such danger is intended to be prevented by electrical emergency switching or by electrical switching for mechanical maintenance of non-electrical parts of such equipment.

Dead: at or about zero voltage in relation to earth, and disconnected from any live system.

Earth: the conductive mass of the Earth, whose electric potential at any point is conventionally taken as zero.

Earthed: connected to Earth through switchgear with an adequately rated earthing capacity or by approved earthing leads.

Electrical equipment: anything used, intended to be used or installed for use to generate, provide, transmit, transform, rectify, convert, conduct, distribute, control, store, measure or use electrical energy.

Electrical installation: an assembly of associated electrical equipment supplied from a common origin to fulfil a specific purpose and having certain coordinated characteristics.

Isolated: the disconnection and separation of the electrical equipment from every source of electrical energy in such a way that this disconnection and separation is secure.

Live: electrically charged.

Notices

    Caution notice: a notice in approved form conveying a warning against interference.

    Danger notice: a notice in approved form reading 'Danger'.

Responsible Director: the director of the company, partner or owner responsible for safety.

Supervisor

  i *immediate supervisor*: a person (having adequate technical knowledge, experience and competence) who is regularly available at the location where work is in progress or who attends the work area as is necessary to ensure the safe performance and completion of work

  ii *personal supervisor*: a person (having adequate technical knowledge, experience and competence) such that he/she is at all times during the course of the work in the presence of the person being supervised.

Voltage: voltage by which an installation (or part of an installation) is designated. The following ranges of nominal voltage (r.m.s. values for a.c.) are defined:

    Extra-low: normally not exceeding 50 V a.c. or 120 V ripple-free d.c., whether between conductors or to earth

    Low: normally exceeding extra-low but not exceeding 1000 V a.c. or 1500 V d.c. between conductors, or 600 V a.c. or 900 V d.c. between conductors and earth. The actual voltage of the installation may differ from the nominal value by a quantity within normal tolerances.

    High Voltage: a voltage exceeding 1000 V a.c. or 1500 V d.c. between conductors, or 600 V a.c. or 900 V d.c. between conductors and earth.

A.1.3 Other safety rules and related procedures

In addition to the application of these safety instructions, other rules and procedures as issued by the employer, or by other authorities, shall be complied with in accordance with management instructions.

In that employees may be required to work in locations, or on or near electrical equipment, installations and appliances, that are not owned or controlled by the employer, these safety instructions have been produced to reasonably ensure safe working, since no other rules/instructions will normally be applicable. However, where the owner has his own rules/instructions and procedures, agreement shall be reached between the company and the owner on which rules/instructions shall be applied. Such agreement shall be made known to the employees concerned.

A.1.4 Information and instruction
Arrangements shall be made to ensure:

i that all employees concerned are adequately informed and instructed as to any equipment, installations or appliances that are associated with work and which legal requirements, safety rules and related procedures shall apply

ii that other persons who are not employees but who may be exposed to danger by the work also receive reasonably adequate information.

A.1.5 Issue of safety instructions
Employees and other persons issued with safety instructions shall sign a receipt for a copy of these safety instructions (and any amendments thereto) and shall keep them in good condition and have them available for reference as necessary when work is being carried out under these safety instructions.

A.1.6 Special procedures
Work on, or test of, equipment, installations and appliances to which rules cannot be applied, or for special reasons should not be applied, shall be carried out in accordance with recognised good practice.

A.1.7 Objections
When any person receives instructions regarding work covered by these safety instructions and objects, on safety grounds, to the carrying out of such instructions, the person issuing them shall have the matter investigated and, if necessary, referred to a higher authority for a decision before proceeding.

A.1.8 Reporting of accidents and dangerous occurrences
All accidents and dangerous occurrences, whether of an electrical nature or not, shall be reported in accordance with *The Reporting of Injuries, Diseases and Dangerous Occurrences Regulations 1995*.

A.1.9 Health and safety
The employer and all employees have a duty to comply with the relevant provisions of the *Health and Safety at Work etc. Act 1974* and with other relevant statutory provisions such as the *Factories Act 1961* and the various Regulations affecting health and safety, including electrical safety. Additionally, authoritative guidance is available from the Health and Safety Executive and other sources.

In addition to these statutory duties and any other responsibilities separately allocated to them, all persons who may be concerned with work shall be conversant with, and comply with, those safety instructions and codes of practice relevant to their duties. Ignorance of legal requirements, or of safety instructions and related procedures, shall not be accepted as an excuse for neglect of duty. If any person has any doubt as to any of these duties he should report the matter to his immediate supervisor.

A.1.10 Compliance with safety instructions
It is the duty of everyone who may be concerned with work covered by these safety instructions, to ensure their implementation and to comply with them and related codes of practice. Ignorance of the relevant legal requirements, safety instructions, codes of practice or approved procedures is not an acceptable excuse for neglect of duty.

The responsibilities placed upon persons may include all or part of those detailed in this section, depending on the role of the persons.

Any written authorisation given to persons to perform their designated role in implementing the safety instructions must indicate the work permitted.

Whether employees are authorised as competent or not, all have the following duties which they must ensure are implemented:

▷ all employees shall comply with these safety instructions when carrying out work, whether instructions are issued orally or in writing.
▷ all employees shall use safe methods of work, safe means of access and the personal protective equipment and clothing provided for their safety.
▷ all employees when in receipt of work instructions shall:
   i be fully conversant with the nature and extent of the work to be done
   ii read the contents and confirm to the person issuing the instructions that they are fully understood
   iii during the course of the work, adhere to, and instruct others under their charge to adhere to, any conditions, instructions or limits specified in the work instructions
   iv when in charge of work, provide immediate or personal supervision as required.

A.2 Non-electrical

A.2.1 Scope
The non-electrical part of the safety instructions shall be applied to work by employees in activities that are non-electrical. This work may involve:

  i work on customers' premises
  ii work on employer's premises
  iii work on the public highway or in other public places.

The safety instructions applicable to this work are those contained in Sections A.1 and A.2 of the safety instructions. When work of an electrical nature is being carried out, all sections (A.1, A.2 and A.3) of the safety instructions apply.

A.2.2 General principle
The general principle is to avoid accidents. Most accidents arise from simple causes and can be prevented by taking care.

A.2.3 Protective clothing and equipment
The wearing of protective clothing and the use of protective equipment can, in appropriate circumstances, considerably reduce the severity of injury should an accident occur.

Where any work under these safety instructions and related procedures takes place, appropriate safety equipment and protective clothing of an approved type shall be issued and used.

At all times employees are expected to wear sensible clothing and footwear having regard to the work being carried out. Further references are made, in particular circumstances, to the use of gloves. Hard hats must be worn at all times when there is a risk of head injury and particularly on construction sites.

Where there is danger from flying particles of metal, concrete or stone, suitable eye protection must be provided and must be used by employees. If necessary, additional screens must be provided to protect other persons in the vicinity.

A.2.4 Good housekeeping

Tidiness, wherever work is carried out, is the foundation of safety; good housekeeping will help to ensure a clean, tidy and safe place of work.

Particular attention should be paid to:

  i picking up dropped articles immediately

  ii wiping up any patches of oil, grease or water as soon as they appear and if necessary, spreading sand or sawdust

  iii removing rubbish and scrap to the appropriate place

  iv preventing objects falling from a height by using containers for hand tools and other loose material

  iv ensuring stairs and exits are kept clear.

No job is completed until all loose gear, tools and materials have been cleared away and the workplace left clean and tidy. Most falls are caused by slippery substances or loose objects on the floor, and good housekeeping will remove most of the hazards that can occur.

A.2.5 Safe access

It is essential that every place of work is at all times provided with safe means of access and exit, and these routes must be maintained in a safe condition.

Keeping the workplace tidy minimises the risk of falling, which is the major cause of accidents, but certain special hazards associated with work in confined spaces require particular attention.

Ladders

All ladders should be of sound construction, uniquely identified and free from apparent defects. This is of particular importance in connection with timber ladders.

The following practices should always be observed:

▸ ladders should be checked before use; any defects must be reported and the equipment clearly marked and not used until repaired

▸ all ladders should be regularly inspected by a competent person and a record kept

▸ ladders in use should stand on a level and firm footing; loose packing should not be used to support the base

▸ ladders should be used at the correct angle, i.e. for every four metres up, the bottom of the ladder must be one metre out

▸ ladders should be lashed at the top when in use, but when this is not practicable they should be held secure at the bottom

▸ the ladder top should extend to a height of at least one metre above any landing place

▸ hand tools and other material should not be carried in the hand when ascending or descending ladders; a bag and sash line should be used

▸ suitable crawling ladders or boards must be used when working on asbestos cement and other fragile roofs. Permanent warning notices should be placed at the means of access to these roofs.

(Note: In a situation where no ladder is available, and the work requires a small step up, it is the employee's responsibility to ensure that any other article used for the purpose is totally suitable.)

Openings in floors

Every floor opening must be guarded, and it is important that other occupants of the workplace are made aware of these hazards. In addition to the risk of persons falling through any opening, there is also a risk from falling objects; safe placing of tools, materials and other objects when working near openings, holes or edges, or at any height, will also prevent accidents.

If work has to be carried out in confined spaces such as tunnels and underground chambers, the atmosphere may be deficient in oxygen or may contain dangerous fumes or substances. *The Electricity Association engineering recommendations, ERG64, safety in cableways* must be followed in these circumstances.

A.2.6 Lighting

Good lighting, whether natural or artificial, is essential to the safety of people whether at the workplace or moving about. If natural lighting is inadequate, it must be supplemented by adequate and suitable artificial lighting. If danger may arise from a power failure, adequate emergency lighting is required.

A.2.7 Lifting and handling

All employees must be trained in the appropriate lifting and handling techniques according to the type of work undertaken (see Chapter 2).

A.2.8 Fire precautions

All employees must be thoroughly conversant with the procedure to be followed in the event of fire. Whether working on customers' premises or elsewhere, employees should familiarise themselves with escape routes, fire precautions etc., before commencing work.

Fire exits must always be kept clear, and access to fire fighting equipment unobstructed.

All fire fighting equipment that is the employer's responsibility must be regularly inspected, maintained and recorded whether by local supervisor, safety supervisor or appropriate third party. Individuals should report any apparent damage to equipment.

A.2.9 Hand tools

Hand tools must be suitable for the purpose for which they are being used and are the responsibility of those using them. They must be maintained in good order and any which are worn or otherwise defective must be reported to a supervisor. Approved insulated tools are available for work on live electrical equipment.

A.2.10 Mechanical handling

Fork-lift and similar trucks must only be driven by operators who have been properly trained, tested and certified for the type of trucks they have to operate. Supervisors should control the issue and return of the truck keys and they should ensure that a daily check of the truck and its controls is carried out by the operator.

A.2.11 Portable power tools

All portable electrical apparatus including cables, portable transformers and other ancillary equipment should be inspected before use and maintained and tested at regular intervals.

Trailing cables are frequently damaged and exposed to wet conditions. Users must report all such damage and other defects as soon as possible, and the faulty equipment must be immediately withdrawn from use.

When not in use power tools should be switched off and disconnected from the source of supply.

A.2.12 Welding, burning and heating processes

General
Welding, burning and heating processes involving the use of gas and electricity demand a high degree of skill and detailed knowledge of the appropriate safety requirements.

▸ specific safety instructions must be issued to employees using such equipment
▸ suitable precautions should be taken, particularly when working overhead, to prevent fire or other injury from falling or flying sparks
▸ all heating, burning and welding equipment must be regularly inspected, and a record kept.

Propane
Propane is a liquefied petroleum gas stored under pressure in cylinders which must be stored vertically in cool, well-ventilated areas, away from combustible material, heat sources and corrosive conditions. Cylinders must be handled carefully and not allowed to fall from a height; when transported, they must always be carried in an approved restraint.

When the cylinder valve is opened the liquid boils, giving off a highly flammable gas. The gas is heavier than air and can give rise to a highly explosive mixture. It is essential, therefore, that valves are turned off after use.

A.2.13 Machinery
All machinery must be guarded as necessary to prevent mechanical hazards.

Facilities shall be provided for isolating and locking off the power to machinery, and work on machines shall not commence unless isolation and locking off from all sources of power has been effected and permits issued to work.

A.3 Electrical

A.3.1 Scope
This part of the Safety Instructions shall be applied to electrical work. This work may involve:

i work on employer's equipment
ii work on customers' electrical installations
iii work on customers' electrical appliances.

This work will normally be concerned with equipment, installations and appliances at low voltages. In the event of work needing to be carried out on high voltage equipment and installations (i.e. where the voltage exceeds 1000 V a.c.), additional instructions and procedures laid down for high voltage work must be issued to those employees who carry out this work.

A.3.2 General principle
As a general principle, and wherever reasonably practicable, work should only be carried out on equipment that is dead and isolated from all sources of supply. Such equipment should be proved dead by means of an approved voltage testing device

which should be tested before and after verification, or by clear evidence of isolation taking account of the possibility of wrong identification or circuit labelling.

Equipment should always be assumed to be live until it is proved dead. This is particularly important where there is a possibility of backfeed from another source of supply.

A.3.3 Information prior to commencement of work

According to the complexity of the installation, the following information may need to be provided before specified work is carried out:

i details of the supply to the premises, and to the system and equipment on which work is to be carried out

ii details of the relevant circuits and equipment and the means of isolation

iii details of any customer's safety rules or procedures that may be applicable to the work

iv the nature of any processes or substances which could give rise to a hazard associated with the work, or other special conditions that could affect the working area, such as the need for special access arrangements

v emergency arrangements on site

vi the name and designation of the person nominated to ensure effective liaison during the course of the work.

Where the available information, or the action to be taken as a result of it, is considered by the person in charge of the work to be inadequate for safe working, such work should not proceed until that inadequacy has been removed or a decision obtained from a person in higher authority. Defects affecting safe working should be reported to the appropriate supervisor.

A.3.4 Precautions to be taken before work commences on dead electrical equipment

In addition to any special precautions to be taken at the site of the work, such as for the presence of hazardous processes or substances, the following electrical precautions should be taken, according to the circumstances, before work commences on dead electrical equipment:

i The electrical equipment should first be properly identified and disconnected from all points of supply by the opening of circuit-breakers, isolating switches, the removal of fuses, links or current-limiting devices, or other suitable means. Approved notices, warning against interference, should be affixed at all points of disconnection.

ii All reasonably practicable steps should be taken to prevent the electrical equipment being made live inadvertently. This may be achieved, according to the circumstances, by taking one or more of the following precautions:

    a approved locks should be used to lock off all switches etc. at points where the electrical equipment and associated circuits can be made live. This should be additional to any lock applied by any other party; the keys to all locks should be retained by the person in charge of the work or in a specially provided key safe,

    b any fuses, links or current-limiting devices involved in the isolation procedures should be retained in the possession of the person in charge of the work, and

    c in the case of portable apparatus, where isolation has been by removal of a plug from a socket-outlet, suitable arrangements should be made to prevent unauthorised re-connection,

d approved notices should be placed at points where the electrical equipment and associated circuits can be made live.

iii The electrical equipment should be proved dead by the proper use of an approved voltage testing device and/or by clear evidence of isolation, such as physically tracing a circuit. Approved testing devices should be checked immediately before and after use to ensure that they are in working order.

iv When work is carried out on timeswitched or other automatically controlled equipment or circuits, the fuses or other means of isolation controlling such equipment or circuits should be removed. On no account should reliance be placed on the timeswitches, limit switches, lock-out push buttons etc., or on any other auxiliary equipment, as means of isolation.

v Where necessary, approved notices should be displayed to indicate any exposed live conductors in the working zone.

vi When it is required to work on dead equipment situated in a substation or similar place where there are exposed live conductors, or adjacent to high voltage plant, the safe working area should be defined by a person authorised in writing under the Safety Rules or under procedures controlling that plant, and all subsequent work must be conducted in accordance with such rules or procedures. Where necessary the exposed live conductors should be adequately screened in an approved manner or other approved means taken to avoid danger from the live conductors.

A.3.5 Precautions to be taken before work commences on or near live equipment

No person shall be engaged in any work activity on or so near any live conductor (other than one suitably covered with insulating material so as to prevent danger) that danger may arise unless:

▶ it is unreasonable in all the circumstances for it to be dead, and
▶ it is reasonable in all the circumstances for them to be at work on or near it while it is live, and
▶ suitable precautions (including where necessary the provision of suitable protective equipment) are taken to prevent injury (regulation 14, *Electricity at Work Regulations*).

Where work is to be carried out on live equipment the following protective equipment should be provided, maintained and used, by adequately trained personnel, in accordance with the safety rules or local procedures as appropriate:

▶ approved screens or screening material
▶ approved insulating standings in the form of hardwood grating or approved rubber insulating mats
▶ approved insulated tools
▶ approved insulating gloves.

When testing, including functional testing or adjustment of electrical equipment, requires covers to be removed so that terminals or connections that are live, or can be made live, are exposed, precautions should be taken to prevent unauthorised approach to or interference with live parts. This may be achieved by keeping the work area under the immediate surveillance of an employee or by erecting a suitable barrier, with approved notices displayed warning against approach and interference. When live terminals or site barriers are being adjusted, only approved insulated tools should be used.

Additional precautions may be required because of the nature of any hazardous process or special circumstances present at the site of the work.

Work on live equipment should only be undertaken where it is unreasonable in all the circumstances for it to be made dead.

A.3.6 Operation of switchgear

The operation of switchgear should only be carried out by a competent person after he has obtained full knowledge and details of the installation and the effects of the intended switching operations.

Under no circumstances must equipment be made operable by hand signals or by a pre-arranged time interval.

A model form of a permit-to-work is included in Figure A.1.

PERMIT-TO-WORK (front)

1. Issue
                                          No..........................

To ....

....

The following apparatus has been made safe in accordance with the Safety Rules for the work detailed on this permit-to-work to proceed:

....

Treat all other apparatus as live

Circuit main earths are applied at

....

Other precautions and information and any local instructions applicable to the work. See notes 1 and 2.

....

The following work is to be carried out

....

....

....

Name (block capitals) ....

Signature ....   Time ....   Date ....

5. Diagram (See over for Sections 2, 3 and 4 of this Permit.)

The diagram should show:
a the safe zone where work is to be carried out
b the points of isolation
c the places where earths have been applied, and
d the locations where 'danger' notices have been posted.

continues

▶ **Figure A.1** Model form of permit-to-work

PERMIT-TO-WORK (back)

2. Receipt
(See Note 2)

No............................

I accept responsibility for carrying out the work on the apparatus detailed on this permit-to-work and no attempt will be made by me, or by the persons under my charge, to work on any other apparatus.

Name (block capitals) ....

Signature ....Time.......................Date...........................

3. Clearance
(See Note 3)

All persons under my charge have been withdrawn and warned that it is no longer safe to work on the apparatus detailed on this permit-to-work, and all additional earths have been removed.

The work is complete/incomplete*

All gear and tools have/have not* been removed

Name (block capitals) ....

Signature ....Time.......................Date...........................

*delete words not applicable

4. Cancellation
(See Note 3)

This permit-to-work is cancelled.

Name (block capitals) ....

Signature ....Time.......................Date...........................

Diagram (continue below if needed)

▶ **Figure A.1** *continued*

Notes on Model Form of Permit-to-Work

Note 1 Access to and work in fire protected areas

Automatic control

Unless alternative approved procedures apply because of special circumstances then before access to, or work or other activities are carried out in, any enclosure protected by automatic fire extinguishing equipment:

a The automatic control shall be rendered inoperative and the equipment left on hand control. A caution notice shall be attached.

b Precautions taken to render the automatic control inoperative and the conditions under which it may be restored shall be noted on any safety document or written instruction issued for access, work or other activity in the protected enclosure.

c The automatic control shall be restored immediately after the persons engaged on the work or other activity have withdrawn from the protected enclosure.

Note 2 Procedure for issue and receipt

a A permit-to-work shall be explained and issued to the person in direct charge of the work, who after reading its contents to the person issuing it, and confirming that he understands it and is conversant with the nature and extent of the work to be done, shall sign its receipt and its duplicate.

b The recipient of a permit-to-work shall be a competent person who shall retain the permit-to-work in his possession at all times while work is being carried out.

c Where more than one working party is involved a permit-to-work shall be issued to the competent person in direct charge of each working party and these shall, where necessary, be cross-referenced one with another.

Note 3 Procedure for clearance and cancellation

a A permit-to-work shall be cleared and cancelled:
   i when work on the apparatus or conductor for which it was issued has been completed;
   ii when it is necessary to change the person in charge of the work detailed on the permit-to-work;
   iii at the discretion of the responsible person when it is necessary to interrupt or suspend the work detailed on the permit-to-work.

b The recipient shall sign the clearance and return to the responsible person who shall cancel it. In all cases the recipient shall indicate in the clearance section whether the work is 'complete' or 'incomplete' and that all gear and tools 'have' or 'have not' been removed.

c Where more than one permit-to-work has been issued for work on apparatus or conductors associated with the same circuit main earths, the controlling engineer shall ensure that all such permits-to-work have been cancelled before the circuit main earths are removed.

Note 4 Procedure for temporary withdrawal or suspension

Where there is a requirement for a permit-to-work to be temporarily withdrawn or suspended this shall be in accordance with an approved procedure.

Reporting of accidents

B

B.1 Introduction

The Health and Safety Executive publishes a guide to the *Reporting of Injuries, Diseases and Dangerous Occurrences Regulations 1995*, reference L73.

The following events are reportable within an appropriate period of time:

▸ when someone at work is unable to do their normal work for more than three days, as a result of an injury caused by an accident at work
▸ on the death of an employee, if this occurs some time after a reportable injury which led to that employee's death, but not more than one year afterwards
▸ a person at work suffers one of a number of specified diseases; specified diseases are set out in a schedule to the regulations.

B.2 Reporting and notifying

In the event of there being a notifiable event the responsible person shall: *Forthwith notify the relevant enforcing authority thereof by the quickest practical measure, and within ten days send a report thereof to the relevant enforcing authority on a form approved for the purpose.*

The responsible person in general is the person at the time having control of the premises at which, or in connection with the work at which, the accident or dangerous occurrence or reportable disease happened.

If you have an accident on your premises, you may not be guilty of any offence but you certainly will be if you do not report it. The *Reporting of Injuries, Diseases and Dangerous Occurrences Regulations 1995* came into force on 1 April 1996. They apply a single set of report requirements to all work activities in Great Britain and in the off-shore oil and gas industries.

The reports will be compiled by the Health and Safety Executive and local authorities. This should provide valuable information as to where the risks are, how they arise and indicate if there are any unfavourable trends. The data can also be useful to guide employers as to how best they might prevent injury and ill health.

B.3 Where the Regulations apply

The Regulations apply in places of work, and to events that arise out of or in connection with work activities, as covered by the *Health and Safety at Work etc. Act*. The Regulations apply to Great Britain but not Northern Ireland, where separate Regulations

© The Institution of Engineering and Technology

are to be made. They also apply to certain work activities carried out in United Kingdom territorial waters. The Regulations apply to mines under the sea and other activities in territorial waters, such as loading and unloading, construction, repair and diving.

B.4 At the accident

If any of the following accidents should occur, the enforcing authority must be notified by the quickest practical means, such as by telephone:

a the death of any person as a result of an accident, whether or not they are at work
b a person suffers a major injury as a result of an accident
c someone who is not at work suffers an injury as a result of an accident arising out of or in connection with work and that person is taken to hospital.

If there is a dangerous occurrence
Dangerous occurrences are accidents or events which have not necessarily resulted in a reportable injury, but have the potential to cause such an injury.

Dangerous occurrences are listed in schedule 2 to the Regulations. As examples of the sort of dangerous occurrences that are reportable and are concerned with electrical matters, the following are included:

▸ *Electrical short-circuit*: electrical short-circuit or overload attended by fire or explosion which results in the stoppage of the plant involved for more than 24 hours, or which has the potential to cause the death of any person.
▸ *Overhead lines*: any unintentional incident in which plant or equipment either:
  comes into contact with an uninsulated overhead electric line of which the voltage exceeds 200 V, or
  causes an electrical discharge from such an electric line by coming into close proximity with it.
▸ *Pressure systems*: the failure of any closed vessel (including a boiler or tube boiler) or any associated pipework, in which the internal pressure was above or below atmospheric pressure, where the failure has the potential to cause the death of any person.
▸ *Lifting machinery etc.*: the collapse or overturning of, or failure of any load-bearing part of any lift or hoist, crane or derrick, mobile power access platform, access cradle or window cleaning cradle, excavator, pile driving frame or fork-lift truck.

B.5 Enforcing authority

The enforcing authority may be either the Health and Safety Executive or the Local Authority. The exact split is determined by the Health and Safety (Enforcing Authority) Regulations 1989. However, Local Authorities are generally responsible for enforcing health and safety legislation in:

▸ retailing
▸ some warehouses
▸ most offices
▸ hotels and catering
▸ sports

▶ leisure
▶ consumer service
▶ places of worship.

If in doubt as to whether there is a need to report, or as to whether it should be reported to the Health and Safety Executive or the local authority, the Health and Safety Executive can be immediately telephoned.

The report forms are available from the Health and Safety Executive. A typical report form is provided in Figure B.1.

B.6 Keeping of records

The responsible person is required to keep a record of:

i Any event that is required to be reported upon.
ii Any disease required to be reported upon.
iii Any such particulars as may be approved by the Health and Safety Executive.

The records are required to be kept either at the place of work to which they relate, or at the usual place of business of the responsible person.

Extracts from records must be sent to the enforcing authority on request and, additionally, an inspector from the enforcing authority may require any part of the records to be produced.

B

Health and Safety at Work etc Act 1974
The Reporting of Injuries, Diseases and Dangerous Occurrences Regulations 1995

Report of an injury or dangerous occurrence

Filling in this form
This form must be filled in by an employer or other responsible person.

Part A

About you

1 What is your full name?

2 What is your job title?

3 What is your telephone number?

About your organisation

4 What is the name of your organisation?

5 What is its address and postcode?

6 What type of work does the organisation do?

Part B

About the incident

1 On what date did the incident happen?

2 At what time did the incident happen?

(Please use the 24-hour clock eg 0600)

3 Did the incident happen at the above address?

Yes ☐ Go to question 4

No ☐ Where did the incident happen?

☐ elsewhere in your organisation – give the name, address and postcode

☐ at someone else's premises – give the name, address and postcode

☐ in a public place – give details of where it happened

If you do not know the postcode, what is the name of the local authority?

4 In which department, or where on the premises, did the incident happen?

F2508 (05.00)

Part C

About the injured person

If you are reporting a dangerous occurrence, go to Part F. If more than one person was injured in the same incident, please attach the details asked for in Part C and Part D for each injured person.

1 What is their full name?

2 What is their home address and postcode?

3 What is their home phone number?

4 How old are they?

5 Are they

☐ male?

☐ female?

6 What is their job title?

7 Was the injured person (tick only one box)

☐ one of your employees?

☐ on a training scheme? Give details:

☐ on work experience?

☐ employed by someone else? Give details of the employer:

☐ self-employed and at work?

☐ a member of the public?

Part D

About the injury

1 What was the injury? (eg fracture, laceration)

2 What part of the body was injured?

Figure B.1 The report form

3 Was the injury (tick the one box that applies)

☐ a fatality?

☐ a major injury or condition? (see accompanying notes)

☐ an injury to an employee or self-employed person which prevented them doing their normal work for more than 3 days?

☐ an injury to a member of the public which meant they had to be taken from the scene of the accident to a hospital for treatment?

4 Did the injured person (tick all the boxes that apply)

☐ become unconscious?

☐ need resuscitation?

☐ remain in hospital for more than 24 hours?

☐ none of the above.

Part E

About the kind of accident

Please tick the one box that best describes what happened, then go to Part G.

☐ Contact with moving machinery or material being machined

☐ Hit by a moving, flying or falling object

☐ Hit by a moving vehicle

☐ Hit something fixed or stationary

☐ Injured while handling, lifting or carrying

☐ Slipped, tripped or fell on the same level

☐ Fell from a height

How high was the fall?

☐ [metres]

☐ Trapped by something collapsing

☐ Drowned or asphyxiated

☐ Exposed to, or in contact with, a harmful substance

☐ Exposed to fire

☐ Exposed to an explosion

☐ Contact with electricity or an electrical discharge

☐ Injured by an animal

☐ Physically assaulted by a person

☐ Another kind of accident (describe it in Part G)

Part F

Dangerous occurrences

Enter the number of the dangerous occurrence you are reporting. (The numbers are given in the Regulations and in the notes which accompany this form)

[   ]

Part G

Describing what happened

Give as much detail as you can. For instance
- the name of any substance involved
- the name and type of any machine involved
- the events that led to the incident
- the part played by any people.

If it was a personal injury, give details of what the person was doing. Describe any action that has since been taken to prevent a similar incident. Use a separate piece of paper if you need to.

Part H

Your signature

Signature

[   ]

Date

[   ]

If returning by post/fax, please ensure this form is signed, alternatively, if returning by E-Mail, please type your name in the signature box

Where to send the form

Incident Contact Centre, Caerphilly Business Centre, Caerphilly Business Park, Caerphilly, CF83 3GG. or email to riddor@natbrit.com or fax to 0845 300 99 24

For official use		
Client number	Location number	Event number
[ ]	[ ]	[ ] ☐ INV REP ☐ Y ☐ N

▷ **Figure B.1** *continued*

Index

© The Institution of Engineering and Technology

Notes